Lecture Notes in Mathematics

A collection of informal reports and seminars
Edited by A. Dold, Heidelberg and B. Eckmann, Zürich

Series: Scuola Normale Superiore, Pisa
Adviser: E. Vesentini

241

Soji Kaneyuki

Scuola Normale Superiore, Pisa/Italia
Nagoya University, Nagoya/Japan

Homogeneous Bounded Domains and Siegel Domains

Springer-Verlag
Berlin · Heidelberg · New York 1971

AMS Subject Classifications (1970): 32-M-10

ISBN 3-540-05702-1 Springer-Verlag Berlin · Heidelberg · New York
ISBN 0-387-05702-1 Springer-Verlag New York · Heidelberg · Berlin

Offsetdruck: Julius Beltz, Hemsbach/Bergstr.

Preface

These notes are based on lectures on "Cayley transforms of homogeneous bounded domains" given at Scuola Normale Superiore, Pisa during the academic year 1970-71.

In § 1, 2 and 3 we give definitions and fundamental results on homogeneous bounded domains, Siegel domains of type II and their affine automorphism groups; Iwasawa subgroups and Iwasawa j-algebras are introduced and considered. In these sections we gave detailed proofs for well-known facts in [14] and [7] .

In §4 we consider the universal and the classifying j-algebras of an Iwasawa j-algebra, due to Piatetski-Sapiro [15] . In §5 we construct the universal domain D and the classifying domain $D(\mathcal{C})$ of a given homogeneous bounded domain D_1 . We discuss the fibring of an arbitrary homogeneous bounded domain coming from a decomposition of its Iwasawa j-algebra into a j-ideal and a j-subalgebra; the fibring of D over $D(\mathcal{C})$ with the standard fibre D_1 is also considered and its universal property is obtained.

The central topics are treated in §6-§9. In §6, we generalize the Borel imbedding in the case of symmetric bounded domains to an arbitrary homogeneous bounded domain D_o ; to do this we need several properties of the full automorphism group of D_o which are proved at the moment by using the main theorem of Vinberg, Gindikin and Piatetski-Sapiro [20] . Contrary to the case of symmetric bounded domains, the complex homogeneous space in which D_o is imbedded holomorphically and equivariantly is not compact in general. As an application of our Borel imbedding, we obtain a holomorphic imbedding of D_o into a complex projective space.

In §7 we obtain various properties of certain subalgebras of the complexification of the Lie algebra of the full automorphism group of a universal domain D . Using these properties and our Borel imbedding, the Cayley transform of D is considered.

In §9, using results in § 7 and 8, it is proved that the Cayley transform of a universal domain D yields a realization of D as a Siegel domain $\mathcal{S}$ of type III, which is a main goal of these notes. From this it follows that the Cayley transform is just a bundle isomorphism of the fibring of D over $D(\mathcal{S})$ onto the natural fibring of $\mathcal{S}$. As an application we obtain more clearly a theorem of Piatetski-Sapiro [15] which suggests that every homogeneous bounded domain D_o is realized as Siegel domain of type III.

In the appendix we give an outline of the proof of the main theorem of Vinberg, Gindikin and Piatetski-Sapiro [20]. Referring the appendix to their original article [20], readers will be able to get the proof of that main theorem.

At the end we remark that universal domains are not symmetric bounded domains in general; but the classical domains of type I, II, and III have structures of universal domains. In these cases our Cayley transforms are different from the partial Cayley transforms due to Wolf-Korányi [21], but both yield the same Siegel domains of type III.

I would like to express my gratitude to Professor E. Vesentini who has encouraged me by his constant interest in these topics. I am also grateful to Miss G. Nannipieri for her excellent preparation of these notes.

Soji Kaneyuki

Pisa, August 2, 1971

This work was partially supported by "Consiglio Nazionale delle Ricerche", Italy.

Table of Contents

§1. The Affine Automorphism Groups of Siegel Domains

Let R be the n-dimensional Euclidean space. A domain V in R is called a convex cone if the following three conditions are satisfied:

Vi) For any $x \in V$ and for any $\lambda > 0$, $\lambda x \in V$.

Vii) If $x, y \in V$, then $x+y \in V$.

Viii) V contains no entire straight lines.

Let W be a complex vector space of complex dimension m. A map F of $W \times W$ into R^c (= the complexification of R) is called a V-hermitian form if the following conditions are satisfied:

Fi) $F(u, v)$ is $\mathbb{C}$-linear in u.

Fii) $\overline{F(u, v)} = F(v, u)$, where the bar means the conjugation of R^c with respect to R.

Fiii) $F(u, u) \in \overline{V}$, where $\overline{V}$ is the closure of V in R.

Fiv) $F(u, u) = 0$ implies $u = 0$.

Then we define the subset $D(V,F)$ of the complex vector space $R^c \times W$ as

$$D(V,F) = \{(x+iy,u) \in R^c \times W \;;\; x, y \in R,\; u \in W,\; y-F(u,u) \in V\}.$$

$D(V,F)$ is a domain in $R^c \times W$, which is called a Siegel domain of type II. If $W=(0)$, the domain $D(V,F)$ is reduced to the tube domain $\{x+iy \in R^c : x \in R, y \in V\}$, which is denoted by $D(V)$ and is called a Siegel domain of type I. A Siegel domain of type I is regarded as a special case of Siegel domains of type II; we do not distinguish type I from type II, unless necessary.

Examples 1) Let $\mathbb{R}^+$ be the set of all positive real numbers. $\mathbb{R}^+$ is a convex cone in $\mathbb{R}$. We have $D(\mathbb{R}^+) = \{x+iy \in \mathbb{C} \;;\; x \in \mathbb{R}, y>0\}$, which is the usual upper half-plane in $\mathbb{C}$.

2) Let $H^+(n,\mathbb{R})$ be the set of all real symmetric positive definite matrices of deg n and $H(n,\mathbb{R})$ be the vector space of all real symmetric matrices of deg n. Then $H^+(n,\mathbb{R})$ is a convex cone in $H(n,\mathbb{R})$. And we have $D(H^+(n,\mathbb{R})) = \{X+iY \;;\; X, Y \in H(n,\mathbb{R}), Y \text{ is positive definite}\}$, which is the Siegel upper half-plane of deg n.

3) Let W be $\mathbb{C}^n$ and for $u=(u_1,\dots,u_n)$, $v=(v_1,\dots,v_n) \in W$

and put $F_o(u,v) = \sum_{i=1}^{n} u_i \bar{v}_i$. Then F_o is a $\mathbb{R}^+$-hermitian form and we have

$$D(\mathbb{R}^+, F_o) = \{(x+iy, u_1, \ldots, u_n) \in \mathbb{C}^{n+1} \; ; \; y - \sum_{i=1}^{n} |u_i|^2 > 0\} .$$

Lemma 1.1. (Piatetski-Sapiro [14]) The domain $D(\mathbb{R}^+, F_o)$ is holomorphically equivalent to the unit open ball $|z_1|^2 + \ldots + |z_{n+1}|^2 < 1$.

Proof. Consider the map ϕ

$$\phi : \begin{cases} z_1 = \dfrac{z-i}{z+i} \\ z_k = \dfrac{2u_{k-1}}{z+i} & 2 \leq k \leq n+1 . \end{cases}$$

Then ϕ is defined on $D(\mathbb{R}^+, F_o)$ and biholomorphic. Moreover we have

$$1 - \sum_{i=1}^{n+1} |z_i|^2 = \frac{4}{|z+i|^2} \left(\operatorname{Im} z - \sum_{i=1}^{n} |u_i|^2\right)$$

which implies that the image of $D(\mathbb{R}^+, F_o)$ under ϕ is the unit open ball.

Proposition 1.2. (Piatetski-Sapiro [14]) A Siegel domain $D(V,F)$ of type II in $R^c \times W$ is holomorphically equivalent to a bounded domain in $R^c \times W$.

Proof. By Viii) , we can choose a linear coordinate $(y_1, \ldots, y_n)$ in R such that $V \subset \{y_1 > 0, \ldots, y_n > 0\}$. And we define the complex coordinate $(z_1, \ldots, z_n)$ by putting $z_\nu = x_\nu + iy_\nu$. With respect to this coordinate $F(u,v)$ can be written as $F(u,v) = (F_1(u,v), \ldots, F_n(u,v))$. By Fii) and Fiii) each $F_i(u,v)$ is a positive semi-definite hermitian form on W . By Fiv)

$$F_1(u,u) = \ldots = F_n(u,u) = 0 \text{ implies } u = 0 \qquad \ldots\ldots(*)$$

Since F_i is a positive semi-definite hermitian form, we can write F_i as

$$F_i(u,u) = |L_{i1}(u)|^2 + \ldots + |L_{is_i}(u)|^2, \qquad 1 \leq i \leq n,$$

where each $L_{ij}(u)$ is a linear form on W. Let us take a subset $\mathcal{L}$ of these $L_{ij}(u)$'s satisfying the following conditions:

1) $\mathcal{L}$ is a linearly independent subset of $\{L_{ij}(u)\}$.
2) Every $L_{ij}(u)$ can be written as a linear combination of elements in $\mathcal{L}$.

The condition (✱) implies that if $L_{ij}(u)=0$ for all i, j, then $u=0$. Hence we can see that the number of elements in $\mathcal{L}$ is equal to $\dim W=m$. Let $\tilde{F}_i(u,u)$ be the hermitian form which is obtained from $F_i(u,u)$ by eliminating all the $L_{ij}(u)$ which do not belong to $\mathcal{L}$. And put

$$\tilde{F}(u,u) = (\tilde{F}_1(u,u),\dots,\tilde{F}_n(u,u)) .$$

It is easy to see that the following inclusions are valid:

$$D(V,F) \subset \left\{ \begin{array}{c} y_1-F_1(u,u)>0 \\ \dots\dots \\ y_n-F_n(u,u)>0 \end{array} \right\} \subset \left\{ \begin{array}{c} y_1-\tilde{F}_1(u,u)>0 \\ \dots\dots \\ y_n-\tilde{F}_n(u,u)>0 \end{array} \right\} .$$

Let $\mathcal{L} = \{L_1(u),\dots, L_m(u)\}$ and put $L_1(u) = u_1',\dots, L_m(u) = u_m'$. Then $(u_1',\dots,u_m')$ is a linear coordinate in W. With respect to this coordinate the domain defined by $y_\nu-\tilde{F}_\nu(u,u)>0$ $(1\le i\le n)$ can be represented as

$$\left\{ \begin{array}{l} y_1 - (|u_1'|^2+\dots+|u_{\ell_1}'|^2) > 0 \\ \qquad\dots\dots \\ y_n - (|u_{\ell_{n-1}+1}'|^2+\dots+|u_m'|^2) > 0 \end{array} \right. .$$

By Lemma 1.1. this domain is holomorphically equivalent to the product of n copies of unit open balls, which proves the proposition for the case where $D(V,F)$ is not of type I. If $D(V,F)$ is of type I, then it is contained in the product of n copies of the usual upper half-planes, which is equivalent to the polydisc of n dimension.

QED

Now let $\mathrm{Aff}(R^c\times W)$ be the group of all complex affine transformations of the vector space $R^c \times W$.

<u>Definition 1.3.</u> For a Siegel domain $D(V,F)$, we put

$$G_a = \{g \in Aff(R^c \times W) \ ; \ g(D(V,F)) = D(V,F)\}.$$

Then G_a is a closed subgroup of $Aff(R^c \times W)$, which is called <u>the affine automorphism group of</u> $D(V,F)$.

We introduce a multiplication in the product $R \times W$ as follows:

$$(a',c')(a,c) = (a'+a-2ImF(c,c'),\ c+c') ,$$

where the imaginary part is taken with respect to R . With this multiplication $R \times W$ becomes a Lie group, which is denoted by RW . In this case the identity of RW is $(0,0)$ and $(a,c)^{-1} = (-a,-c)$. Let $(a,c) \in RW$. Let us define the transformation $g_{(a,c)}$ on $R^c \times W$:

$$g_{(a,c)}(x+iy,u) = (x+a+i(y+2F(u,c)+F(c,c)),\ u+c)$$

Obviously $g_{(a,c)} \in Aff(R^c \times W)$. The map $\phi : (a,c) \mapsto g_{(a,c)}$ is a representation of RW into $Aff(R^c \times W)$. ϕ is faithful. In fact, suppose that $g_{(a,c)}$ is the identity. Then $(0,0) = g_{(a,c)}(0,0) = (a+iF(c,c),c)$, which implies that $a=c=0$. Moreover we can see that $\phi(RW)$ is a closed subgroup of $Aff(R^c \times W)$. From now on the group RW will be identified with $\phi(RW)$.

<u>Definition 1.4.</u> For a Siegel domain $D(V,F)$ we define a subset S of $R^c \times W$ by putting:

$$S = \{(x+iy,u) \in R^c \times W \ : \ y-F(u,u)=0\}.$$

<u>Lemma 1.5.</u> <u>The group</u> RW <u>is a subgroup of</u> G_a <u>and acts on</u> S <u>simply transitively</u>.

<u>Proof.</u> The polynomial $y-F(u,u)$ is invariant under the action of RW. In fact, let $(a,c) \in RW$ and let $(x'+iy',u') = (a,c)(x+iy,u)$. Then

$$y'-F(u',u') = \operatorname{Im}\{x+a+i(y+2F(u,c)+F(c,c))\}-F(u+c,u+c)$$

$$= y+2\operatorname{Re} F(u,c)+F(c,c)-F(u,u)-F(u,c)-F(c,u)-F(c,c)$$

$$= y-F(u,u) \ ,$$

which implies that RW is a subgroup of G_a and that S is stable under the action of RW . Let $(x+iy,u)\varepsilon S$. Then $(-x,-u)\varepsilon RW$ carries $(x+iy,u)$ to $(0,0)\varepsilon S$. Hence RW is transitive. Moreover we can see immediately that the isotropy subgroup of RW at (0,0) is reduced to the identity.

QED

Let $D = D(V,F) \subset R^c\times W$ be a Siegel domain of type II. Let $(z_1,\ldots,z_n)$ be the same coordinate as in the proof of Proposition 1.2, and let $(u_1,\ldots,u_m)$ be a linear coordinate in W .

__Definition 1.6.__ Let E be the set of all holomorphic functions on D such that

1) f is holomorphic on the closure $\bar{D}$, that is to say, f is holomorphic on some open set containing $\bar{D}$.
2) $|f(z,u)| \to 0$ as $(z,u)\varepsilon\bar{D}$ and $\sum_{i=1}^{n} |z_i|^2+ \sum_{i=1}^{m} |u_i|^2 \to \infty$.

For every $f\varepsilon E$, $\sup_{\bar{D}}|f|$ is attained at a point in $\bar{D}$. From the straight-forward computations we get

__Lemma 1.7.__ ([8]) __Let__

$$f^*(z,u) = \prod_{k=1}^{n} \frac{1}{z_k+i} \ .$$

__Then__ $f^*\varepsilon E$ __and__ $\sup_{\bar{D}}|f^*|$ __is attained only at__ $(0,0)\varepsilon S$.

__Lemma 1.8.__ __For each point__ $p_o\varepsilon S$, __there exists a function__ $f\varepsilon E$ __such that__ $\sup_{\bar{D}}|f|$ __is attained only at__ p_o .

__Proof.__ It can be seen immediately that the group RW acts on the family E in the natural manner. By Lemma 1.5., there exists $g \varepsilon RW$ which carries p_o to $(0,0)$. Then the holomorphic function $f=f^{*} \circ g$ belongs to E . Furthermore $g\bar{D}=\bar{D}$ holds, since $g \varepsilon G_a$ (by Lemma 1.5.) and g is a homeomorphism of $R^c \times W$. Hence by Lemma 1.7. we see that $\sup_{\bar{D}}|f|$ is attained only at $g^{-1}(0,0)=p_o$.

<u>Definition 1.9.</u> For a Siegel domain $D(V,F) \subset R^c \times W$, we have defined the set S . We define the subgroup $G(S)$ by putting

$$G(S) = \{g \varepsilon \mathrm{Aff}(R^c \times W) \ ; \ gS=S\} \ .$$

$G(S)$ is a closed subgroup of $\mathrm{Aff}(R^c \times W)$.

<u>Proposition 1.10.</u> G_a <u>is a subgroup of</u> $G(S)$.

<u>Proof.</u> For each $p \varepsilon S$ and $g \varepsilon G_a$, it is enough to prove $g \cdot p \varepsilon S$. Let H^+ be the closed upper half-plane in $\mathbb{C}$, that is to say , $H^+ = \{t \varepsilon \mathbb{C} \ ; \ \mathrm{Im}t \geq 0\}$. Let $g \cdot p=q=(x_o+iy_o,u_o)$. Then by the same reason as in the proof of Lemma 1.8., $q \varepsilon \bar{D}$. Let us define the holomorphic map α of H^+ into $R^c \times W$ as

$$\alpha(t) = \big(x_o+iF(u_o,u_o)+t(y_o-F(u_o,u_o)),u_o\big) \ .$$

First we need to show that $\alpha(t) \varepsilon \bar{D}$ for all $t \varepsilon H^+$. Note that $\bar{D} = \{(x+iy,u) \ ; \ y-F(u,u) \varepsilon \bar{V}\}$. Since $q \varepsilon \bar{D}$, we have $y_o-F(u_o,u_o) \varepsilon \bar{V}$. On the other hand

$$\mathrm{Im}\big(x_o+iF(u_o,u_o)+t(y_o-F(u_o,u_o))\big)-F(u_o,u_o) = (\mathrm{Im}t)\big(y_o-F(u_o,u_o)\big) \ .$$

Furthermore, since $\mathrm{Im}t \geq 0$ and $\bar{V}$ is a closed cone, $(\mathrm{Im}t)\big(y_o-F(u_o,u_o)\big) \varepsilon \bar{V}$, which implies that $\alpha(t) \varepsilon \bar{D}$. Next let us define the holomorphic map β of H^+ into $R^c \times W$ as $\beta(t) = g^{-1}\alpha(t)$. Then $\beta(H^+) = g^{-1}\alpha(H^+) \subset g^{-1}\bar{D} = \bar{D}$.

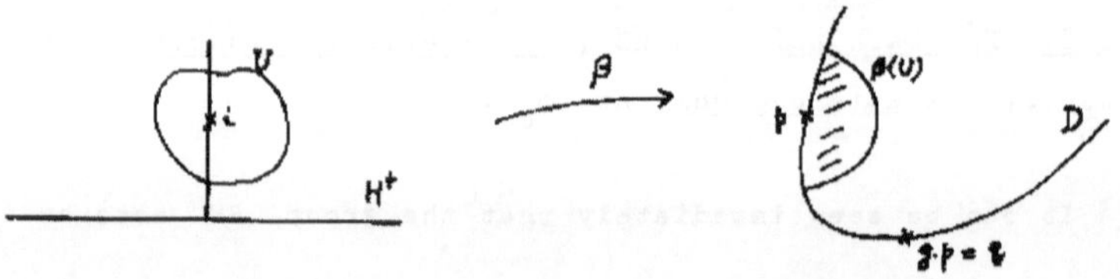

We have $\beta(i)=p$. In fact, $\beta(i)=g^{-1}\alpha(i)=g^{-1}(x_o+iy_o,u_o)=g^{-1}q=p$. By Lemma 1.8., we can choose $f\epsilon E$ such that $\sup_{\overline{D}}|f|$ is attained only at $p\epsilon S$. If we choose a small neighborhood U of i in H^+ , then $f\circ\beta$ is holomorphic on U and $\sup_U|f\circ\beta|$ is attained at i , which is an interior point of U . Hence by the maximal modulus principle we see that $f\circ\beta$ is a constant on U . Therefore, by the property of f , β is a constant on U , which implies that α is a constant on U . Let us take two points $ia_1, ia_2\epsilon U$, where a_1, a_2 are real . Then

$$0 = \alpha(ia_1)-\alpha(ia_2) = \big(i(a_1-a_2)(y_o-F(u_o,u_o)),0\big) ,$$

which implies that $y_o-F(u_o,u_o)=0$. So $q\epsilon S$.

QED

<u>Proposition 1.11.</u> (Piatetski-Sapiro [14]) <u>Let</u> $g\epsilon Aff(R^c\times W)$. <u>Then</u> $g\epsilon G(S)$ <u>if and only if</u> g <u>can be written in the form</u>

$$g(z,u) = \big(Az+a+2iF(Bu,b)+iF(b,b),Bu+b\big) , \tag{1}$$

<u>where</u> $a\epsilon R$, $b\epsilon W$, $A\epsilon GL(R)$, $B\epsilon GL(W)$ <u>and</u>

$$AF(u,v) = F(Bu,Bv) \quad u,v\epsilon W . \tag{2}$$

<u>Proof.</u> Since g is a complex affine transformation, g can be written as

$$g\begin{pmatrix} z \\ u \end{pmatrix} = \begin{pmatrix} A & A' \\ B' & B \end{pmatrix}\begin{pmatrix} z \\ u \end{pmatrix} + \begin{pmatrix} c \\ b \end{pmatrix} \tag{3}$$

where A, A', B, B' are complex matrices and $c\epsilon R^c$, $b\epsilon W$.

Suppose that $g\epsilon G(S)$. Then $g(0,0)=(c,b)\epsilon S$, since $(0,0)\epsilon S$. Hence we have

$$\mathrm{Im}\, c = F(b,b) \tag{4}$$

Consider the transformation $g_1=(-\mathrm{Re}\,c,-b)\epsilon RW$. g_1 belongs to $G(S)$ by Lemma 1.5. and Proposition 1.10. It is easy to see that

$$g_1g(z,u) = \left(Az-2iF(B'z,b)+A'u-2iF(Bu,b),\ B'z+Bu\right) .$$

We define $\mathbb{C}$-linear maps A_1 and A_2 as

$$\begin{cases} z \mapsto A_1z = Az-2iF(B'z,b) & (5) \\ u \mapsto A_2u = A'u-2iF(Bu,b) . & (6) \end{cases}$$

Then

$$g_1g(z,u) = (A_1z+A_2u,\ B'z+Bu) .$$

Take $(x,0)\varepsilon S$, where $x\varepsilon R$. Since $g_1g\varepsilon G(S)$, we have $g_1g(x,0) = (A_1x,B'x)\varepsilon S$, which implies that $(\mathrm{Im}A_1)x = \mathrm{Im}A_1x = F(B'x,B'x)$ for all $x\varepsilon R$. And the left-hand side is linear in x , while the right-hand side is a polynomial of degree 2 in x . Hence $(\mathrm{Im}A_1)x = F(B'x,B'x) = 0$ for all $x\varepsilon R$. So we have

$$\mathrm{Im}A_1 = 0 ,$$
$$B' = 0 .$$

Therefore by (5) we see that

$$A_1 = A .$$

In particular $A\varepsilon GL(R)$ and $B\varepsilon GL(W)$. And we have

$$g_1g(z,u) = (Az+A_2u,Bu) .$$

Take a point $(iF(u,u),u)\varepsilon S$. Then $g_1g(iF(u,u),u)=(iAF(u,u)+A_2u,Bu)\varepsilon S$. Hence $\mathrm{Im}(iAF(u,u)+A_2u)=F(Bu,Bu)$ is valid, or

$$AF(u,u)+\mathrm{Im}A_2u = F(Bu,Bu) . \tag{7}$$

Take a point $(iF(u,u),\ e^{i\theta}u)\varepsilon S$, where $\theta\varepsilon\mathbb{R}$. The same arguments show that $AF(u,u)+\mathrm{Im}(e^{i\theta}A_2u)=F(Bu,Bu)$. Hence it follows that

$$\mathrm{Im}A_2u = \mathrm{Im}(e^{i\theta}A_2u) \quad \text{for all} \quad \theta\varepsilon\mathbb{R} ,$$

which implies that $A_2u=0$ for all $u\varepsilon W$. So we have $A_2=0$. Therefore it follows from (7) that

$$AF(u,u) = F(Bu,Bu) .$$

Hence, we get (2). Since $A_2=0$, we have from (6)

$$A'u = 2iF(Bu,b)$$

and we see that

$$g(z,u) = (Az+A'u+c,\ Bu+b) = (Az+2iF(Bu,b)+\mathrm{Re}\,c+i\,\mathrm{Im}\,c,\ Bu+b) .$$

Taking (4) into account and putting $\mathrm{Re}\,c=a$, we have (1).

Conversely if g can be written in the form (1) and satisfies (2), then it can be seen that $gS=S$.

QED

<u>Corollary 1.12.</u> <u>Let</u> $g\in G(S)$. <u>Then</u> $g\in G_a$ <u>if and only if</u> A <u>leaves the cone</u> V <u>stable, where</u> A <u>is the same one as in</u> (1).

<u>Proof.</u> Let $y\in V$. Then $(iy,0)\in D(V,F)$. Suppose that $g\in G_a$. Then $g(iy,0)=(iAy+a+iF(b,b),b)\in D(V,F)$. Hence $Ay=\mathrm{Im}\{iAy+a+iF(b,b)\}-F(b,b)\in V$. The converse is easily seen.

From Proposition 1.11. and Corollary 1.12. we have the following theorem of Piatetski-Sapiro [14].

<u>Theorem 1.13.</u> <u>Let</u> $D(V,F)\subset R^c\times W$ <u>be a Siegel domain of type</u> II <u>and</u> G_a <u>the affine automorphism group of</u> $D(V,F)$ <u>and</u> H <u>be the isotropy subgroup of</u> G_a <u>at</u> $(0,0)\in R^c\times W$. <u>Let</u> $(ir,0)\in D(V,F)$, <u>where</u> $r\in R$, <u>and let</u> K_a <u>be the isotropy subgroup of</u> G_a <u>at</u> $(ir,0)$. <u>Then we have</u>

1) $G_a = H\cdot RW$ (semi-direct)

2) $K_a\subset H$

3) <u>There exists a representation</u> ρ <u>of</u> H <u>on</u> R <u>and a representation</u> σ <u>of</u> H <u>on</u> W <u>such that</u>

$$\rho(h)F(u,v) = F(\sigma(h)u,\ \sigma(h)v) \qquad u,\ v\in W \quad h\in H .$$

4) <u>Each element</u> $g=(h,a,c)\in G_a=H\cdot RW$ $(h\in H,\ a\in R,\ c\in W)$ <u>acts on</u> $R^c\times W$ <u>as follows</u>:

$$(h,a,c)(x+iy,u) = (\rho(h)x+a+i(\rho(h)y+2F(\sigma(h)u,c)+F(c,c)),\ \sigma(h)u+c)$$

5) <u>The multiplication in</u> G_a <u>is given by</u>

$$(h',a',c')(h,a,c) = \left(h'h,\ \rho(h')a+a'-2\mathrm{Im}F(\sigma(h')c,c'),\ \sigma(h')c+c'\right).$$

<u>Proof.</u> (1) RW is a subgroup of G_a, and RW and G_a are closed subgroups of $\mathrm{Aff}(R^c\times W)$. So RW is closed in G_a. H is also closed in G_a. Furthermore $H \cap RW=(1)$. In fact, let $g\varepsilon RW \cap H$. Then $g\cdot(0,0)=(0,0)\varepsilon S$. On the other hand g acts on S simply transitively. So $g\cdot(0,0)=(0,0)$ implies that g is the identity.

(2) Let $g\varepsilon K_a$ and let us write g in the form (1) in Proposition 1.11. Then we have $(ir,0)=g(ir,0)=(iAr+a+iF(b,b),b)$. Thus $a=b=0$. Hence it follows that $g\varepsilon H$.

(3) Let $g\varepsilon G_a$. The action of g on $R^c\times W$ is given by (1) in Proposition 1.11. Hence it follows that $g\varepsilon H$ if and only if $g(z,u) = (Az,Bu)$. If we define ρ and σ as $\rho(g)=A$ and $\sigma(g)=B$, then ρ and σ satisfy the assertion in 3) (cf. (2) in Proposition 1.11.).

(4) This is the version of Proposition 1.11.

(5) is obtained by direct calculations from (4).

<u>Example.</u> Let $R=H(2,\mathbb{R})$ and consider the convex cone $H^+(2,\mathbb{R})$ in $H(2,\mathbb{R})$. Let $W=\mathbb{C}^2$. For $u=\binom{u_1}{u_2}$, $v=\binom{v_1}{v_2}\varepsilon\mathbb{C}^2$, we define a $H^+(2,\mathbb{R})$-hermitian form F by putting

$$F(u,v) = \frac{1}{2}(u\,{}^t\bar{v}+\bar{v}\,{}^tu) = \begin{pmatrix} u_1\bar{v}_1 & \dfrac{u_1\bar{v}_2+u_2\bar{v}_1}{2} \\ \dfrac{u_1\bar{v}_2+u_2\bar{v}_1}{2} & u_2\bar{v}_2 \end{pmatrix}.$$

Thus the Siegel domain $D(H^+(2,\mathbb{R}),F)$ is given by

$$D(H^+(2,\mathbb{R}),F) = \left\{(z,u) = \left(\begin{pmatrix} z_1 & z_3 \\ z_3 & z_2\end{pmatrix}, \begin{pmatrix} u_1 \\ u_2\end{pmatrix}\right) \varepsilon\mathbb{C}^5 \ :\ \frac{z-{}^t\bar{z}}{2i} - F(u,u) > 0\right\}$$

$$= \left\{(z_1,z_2,z_3,u_1,u_2)\ ;\ \begin{pmatrix} y_1 & y_3 \\ y_3 & y_2\end{pmatrix} - \begin{pmatrix} |u_1|^2 & \mathrm{Re}\,u_1\bar{u}_2 \\ \mathrm{Re}\,u_1\bar{u}_2 & |u_2|^2\end{pmatrix} > 0\right\}$$

$$= \left\{(z_1,z_2,z_3,u_1,u_2)\ ;\ \begin{matrix} y_1-|u_1|^2 > 0 \\ (y_1-|u_1|^2)(y_2-|u_2|^2) - (y_3-\mathrm{Re}\,u_1\bar{u}_2)^2 > 0\end{matrix}\right\},$$

where $y_i = \mathrm{Im}\ z_i$ $(i=1,2,3)$. And we can see that $H = GL(2,\mathbb{R}) \times SO(2)$. Hence we have

$$G_a = \big(GL(2,\mathbb{R}) \times SO(2)\big) \cdot \big(H(2,\mathbb{R})\mathbb{C}^2\big)$$

and $\dim G_a = 12$. Furthermore G_a acts on $D\big(H^+(2,\mathbb{R}),F\big)$ transitively. The isotropy subgroup K_a of G_a at the point $\big(i\begin{pmatrix}1&0\\0&1\end{pmatrix},0\big)=(i,i,0,0,0)$ is $O(2)\times SO(2)$. If we write $h \in H$ in the form $h=\big(\begin{pmatrix}a&b\\c&d\end{pmatrix},e^{i\theta}\big)$, where $\begin{pmatrix}a&b\\c&d\end{pmatrix} \in GL(2,\mathbb{R}), e^{i\theta} \in SO(2)$, then we can see that

$$\rho(h) = \begin{pmatrix}a&b\\c&d\end{pmatrix}$$

$$\sigma(h) = \begin{pmatrix}a&b\\c&d\end{pmatrix}\begin{pmatrix}e^{i\theta}&0\\0&e^{i\theta}\end{pmatrix} \in GL(2,\mathbb{C}) .$$

And for $y=\begin{pmatrix}y_1&y_3\\y_3&y_2\end{pmatrix}$, $u=\begin{pmatrix}u_1\\u_2\end{pmatrix}$, we have

$$\rho(h)y = \begin{pmatrix}a&b\\c&d\end{pmatrix}\begin{pmatrix}y_1&y_3\\y_3&y_2\end{pmatrix}{}^t\begin{pmatrix}a&b\\c&d\end{pmatrix}$$

$$\sigma(h)u = \begin{pmatrix}a&b\\c&d\end{pmatrix}\begin{pmatrix}e^{i\theta}&0\\0&e^{i\theta}\end{pmatrix}\begin{pmatrix}u_1\\u_2\end{pmatrix} .$$

We need some definitions for later considerations.

Definition 1.14. Let $D(V,F)$ be a Siegel domain of type II. If the affine automorphism group G_a acts on $D(V,F)$ transitively, then $D(V,F)$ is called affine homogeneous.

Definition 1.15. Let V be a convex cone in R . Let $\mathrm{Aut}\ V$ be the group of all non-singular linear transformations g of R satisfying $gV=V$. V is called homogeneous if the group $\mathrm{Aut}\ V$ acts transitively on V .

Using theorem 1.13. we can easily see that if $D(V,F)$ is affine homogeneous, then the group $\rho(H)$ acts transitively on V .

§2. The Iwasawa Subgroups

Let D be a bounded domain in the complex n-space $\mathbb{C}^n$, and $G_h(D)$ be the group of all holomorphic automorphisms of D. By the well-known theorem of H. Cartan $G_h(D)$ is a Lie transformation group of D with respect to the compact open topology. D is called homogeneous, if the group $G_h(D)$ acts transitively on D. We know the following theorem of Vinberg, Gindikin and Piatetski-Sapiro [20].

Theorem A. Let D be a homogeneous bounded domain in $\mathbb{C}^n$. Then there exists an affine homogeneous Siegel domain $D(V,F)$ of type II holomorphically equivalent to D.

Hence, to consider homogeneous bounded domains, it is enough to consider affine homogeneous Siegel domains of type II. Throughout this section we will assume that Siegel domains of type II are affine homogeneous.

Let $D(V,F) \subset R^c \times W$ be a (affine homogeneous) Siegel domain of type II. We choose suitable bases in R and W, respectively. Then $R^c \times W$ can be identified with $\mathbb{C}^{n+m}$, where $n=\dim R$ and $m=\dim_c W$. Since $Aff(R^c \times W)$ is represented as a semi-direct product $GL(n+m,\mathbb{C}) \cdot \mathbb{C}^{n+m}$, we can write each element $g \in Aff(R^c \times W)$ in the form $g=(A,a)$, where $A \in GL(n+m,\mathbb{C})$ and $a \in \mathbb{C}^{n+m}$. A is called the linear part of g and a is called the translation part of g. The map which carries $g=(A,a)$ to the matrix $\begin{pmatrix} A & a \\ 0 & 1 \end{pmatrix} \in GL(n+m+1,\mathbb{C})$ is a faithful representation of $Aff(R^c \times W)$. Furthermore $GL(n+m+1,\mathbb{C})$ is canonically and isomorphically imbedded in $GL(2n+2m+2,\mathbb{R})$. From now on we will regard $Aff(R^c \times W)$ as a (Lie) subgroup of $GL(2n+2m+2,\mathbb{R})$.

Definition 2.1. A subgroup G of $Aff(R^c \times W)$ is called real algebraic, if G is an algebraic subgroup of $GL(2n+2m+2,\mathbb{R})$.

This definition is independent of choice of bases in R and W. $Aff(R^c \times W)$ is itself real algebraic.

Let X be a vector space over $\mathbb{R}$. A Lie subgroup G' of $GL(X)$ is called $\mathbb{R}$-triangular if there exists a base of X relative to which each element of G' can be written as an upper triangular matrix. Let G be a linear Lie group. A subgroup G' of G is called a maximal $\mathbb{R}$-triangular if

1) G' is $\mathbb{R}$-triangular and connected

2) A connected $\mathbb{R}$-triangular subgroup G'' of G such that $G'' \supset G'$ coincides with G'.

We will recall several results of Vinberg [19] about $\mathbb{R}$-triangular groups. A group G' is $\mathbb{R}$-triangular if and only if every eigenvalue of each element of G' is real. If G is the identity component of a real algebraic group, then G is represented as the semi-direct product of a maximal compact subgroup of G and a maximal $\mathbb{R}$-triangular subgroup of G. Two maximal $\mathbb{R}$-triangular subgroups of a linear Lie group G are conjugate with respect to an inner automorphism of G.

<u>Definition 2.2.</u> A Lie subgroup G of $Aff(R^c \times W)$ is called <u>$\mathbb{R}$-triangular</u> if G is $\mathbb{R}$-triangular in $GL(2n+2m+2,\mathbb{R})$.

<u>Proposition 2.3.</u> <u>If</u> $D(V,F)$ <u>is affine homogeneous, then</u> G_a° <u>is the identity component of a real algebraic group, where</u> G_a° <u>is the identity component of the affine automorphism group</u> G_a <u>of</u> $D(V,F)$.

<u>Proof.</u> Let N be the normalizer of G_a° in $Aff(R^c \times W)$, and N° be the identity component of N. First we want to prove $N^\circ = G_a^\circ$. Let $p_o \in D(V,F)$, and U_1 be a neighborhood of p_o in $D(V,F)$. Then $U = \{g \in Aff(R^c \times W) : g \cdot p_o \in U_1\}$ is a neighborhood of the identity of $Aff(R^c \times W)$. Take an arbitrary element $g \in U \cap N$. Then we have $gD(V,F) = gG_a^\circ \cdot p_o = G_a^\circ \cdot g \cdot p_o = D(V,F)$. (Note that G_a° is transitive on $D(V,F)$). Hence $g \in G_a \cap U$, which implies that $U \cap N = G_a \cap U$. On the other hand $U \cap N$ (resp. $G_a \cap U$) is a neighborhood of the identity of N (resp. G_a), since N and G_a are closed subgroups. Hence we get $N^\circ = G_a^\circ$.

Next we want to show that N is real algebraic. Let $\mathfrak{g}_a$ be the Lie algebra of G_a. Then by the definition of N we see that N consists of all the elements $g \in Aff(R^c \times W)$ such that $(Adg)\,\mathfrak{g}_a = \mathfrak{g}_a$. Hence if we choose a suitable base in the Lie algebra of $Aff(R^c \times W)$, then we see that

$$N = \{g \in Aff(R^c \times W) \ ; \ Adg = \begin{pmatrix} Ad_{\mathfrak{g}_a} g & * \\ 0 & * \end{pmatrix} \}.$$

And all coefficients of the matrix Adg are rational functions of coefficients of g. Hence the above equality shows that N is defined as the set of zeroes of polynomials of real coefficients in $\mathrm{Aff}(R^c \times W)$, which implies that N is an algebraic subgroup in $GL(2n+2m+2, \mathbb{R})$.

<u>Corollary 2.4. The identity component</u> H° <u>of</u> H <u>coincides with the identity component of a real algebraic group.</u>

<u>Proof.</u> By the definition of H we have $H = G_a \cap GL(n+m, \mathbb{C})$ in $\mathrm{Aff}(R^c \times W)$. Hence by Proposition 2.3. we have $H^\circ = (N \cap GL(n+m, \mathbb{C}))^\circ$. And N and $GL(n+m, \mathbb{C})$ are real algebraic in $\mathrm{Aff}(R^c \times W)$, which implies that $N \cap GL(n+m, \mathbb{C})$ is real algebraic.

<u>Lemma 2.5. An affine homogeneous Siegel domain</u> $D(V,F) \subset R^c \times W$ <u>of type II is a cell.</u>

<u>Proof.</u> As is remarked at the end of §1, if $D(V,F)$ is affine homogeneous, the cone V is homogeneous. Consider the map

$$D(V,F) \ni (x+iy, u) \mapsto (x, y-F(u,u), u) \in R \times V \times W .$$

This is a diffeomorphism of $D(V,F)$ onto $R \times V \times W$. Furthermore V is a cell (Vinberg [18]), which proves the lemma.

<u>Lemma 2.6.</u> $D(V,F)$ <u>can be represented as a coset space</u> G_a° / K_a°, <u>where</u> K_a° <u>is the identity component of</u> K_a. <u>And</u> K_a° <u>is a maximal compact subgroup of</u> G_a°.

<u>Proof.</u> Let K_a^* be the isotropy subgroup of G_a° at $(ir, 0) \in D(V,F)$. Since G_a° is transitive on $D(V,F)$, $D(V,F)$ can be written as the coset space G_a° / K_a^*. Furthermore we have $K_a^* = G_a^\circ \cap K_a$. Hence the identity component of K_a^* coincides with K_a°. So G_a° / K_a° is a covering space of G_a° / K_a^*, which is a cell. Hence we see that $K_a^* = K_a^\circ$. And K_a° is a maximal compact subgroup of G_a°, since G_a° / K_a° is a cell.

QED

The group K_a° is a maximal compact subgroup of G_a° contained in H° (cf. Theorem 1.13.). So K_a° is a maximal compact subgroup of H°. Hence it follows from Corollary 2.4 that

$$H^o = K_a^o \cdot T_1 \quad \text{(semi-direct)}$$

where T_1 is a maximal $\mathbb{R}$-triangular subgroup of H^o .

Definition 2.7. Let D be a homogeneous bounded domain in $\mathbb{C}^n$ and $D(V,F)$ be an affine homogeneous Siegel domain of type II holomorphically equivalent to D . We define a Lie subgroup T of G_a^o by putting

$$T = T_1 \cdot RW \quad \text{(semi-direct)}$$

where the semi-direct product is taken in $G_a = H \cdot RW$. The subgroup T is called an Iwasawa subgroup of $G_h(D)$ with respect to the realization $D(V,F)$.

From the definition we have

$$G_a^o = K_a^o \cdot T \quad \text{(semi-direct)} .$$

T is a simply connected Lie group acting simply transitively on D and $D(V,F)$.

Proposition 2.8. The group T is a maximal $\mathbb{R}$-triangular subgroup of G_a^o .

Proof. First we will prove that T is $\mathbb{R}$-triangular.
Let $g=(A,a)\in Aff(R^c \times W)$, where A (resp. a) is the linear (resp. translation) part of g . If we consider g as an element in $GL(2n+2m+2,\mathbb{R})$, then g can be represented as the matrix

$$\begin{pmatrix} B & -C & b & -c \\ C & B & c & b \\ 0 & 0 & 1 & 0 \\ 0 & 0 & 0 & 1 \end{pmatrix},$$

where $A=B+iC$, $a=b+ic$ and B,C are real matrices and b,c are real vectors. Hence, in order to see whether all eigenvalues of the above matrix are real or not, it is enough to see whether all eigenvalues of the submatrix $\begin{pmatrix} B & -C \\ C & B \end{pmatrix}$ are real or not. This submatrix corresponds to the linear part of g .

Now let $(h,a,c)\in T=T_1\cdot RW$, where $h\in T_1$, $a\in R$, $c\in W$. The linear part of (h,a,c) is given by the matrix

$$\begin{pmatrix} \rho(h) & 0 & * & * \\ 0 & \rho(h) & * & * \\ 0 & 0 & \sigma(h)_1 & -\sigma(h)_2 \\ 0 & 0 & \sigma(h)_2 & \sigma(h)_1 \end{pmatrix} \qquad (\#)$$

where $\sigma(h)_1$ (resp. $\sigma(h)_2$) is real (resp. imaginary) part of $\sigma(h)$. On the other hand $(h,0,0)\in T_1$ is given by

$$\begin{pmatrix} \rho(h) & 0 & 0 & 0 \\ 0 & \rho(h) & 0 & 0 \\ 0 & 0 & \sigma(h)_1 & -\sigma(h)_2 \\ 0 & 0 & \sigma(h)_2 & \sigma(h)_1 \end{pmatrix}.$$

Since T_1 is $\mathbb{R}$-triangular, this matrix has only real eigenvalues. Hence the matrix (#) has only real eigenvalues, which implies that T is $\mathbb{R}$-triangular.

Next we want to show that T is maximal. Let T' be a connected $\mathbb{R}$-triangular subgroup of G_a^o containing T as a proper subgroup. Then $T'\cap K_a^o\neq(1)$, since $G_a^o=K_a^o\cdot T$. Take $t\in T'\cap K_a^o$, which is not the identity. Since T' is connected, we can write t as

$$t = \exp X_1\cdot\exp X_2\cdot\ldots\cdot\exp X_s ,$$

where $X_1,\ldots,X_s$ are elements of the Lie algebra $\mathfrak{t}'$ of T'. On the other hand, if we choose a suitable base in $\mathbb{R}^{2n+2m+2}$, then $\mathfrak{t}'$ is a subalgebra of the Lie algebra of all upper triangular matrices in $\mathfrak{gl}(2n+2m+2,\mathbb{R})$. Hence each X_i has only real eigenvalues, which implies that $\exp X_i$ has only positive eigenvalues. Consequently every eigenvalue λ of t is positive. On the other hand, since $t\in K_a^o$ we have $|\lambda|=1$, from which it follows that all eigenvalues of t are equal to 1. Furthermore t is a semi-simple operator on $\mathbb{R}^{2n+2m+2}$. Hence we see that t is the identity, which is a contradiction. So T is maximal $\mathbb{R}$-triangular.

Remark. Let T_1' be another maximal $\mathbb{R}$-triangular subgroup of H^o, and let us define the group T' as $T'=T_1'\cdot RW$. Then by the same arguments as above, T' is a maximal $\mathbb{K}$-triangular subgroup of G_a^o. Hence T' is conjugate to T in G_a^o. In this sense T is uniquely determined by $D(V,F)$.

§3. j-Algebras

Definition 3.1. Let $\mathfrak{g}$ be a Lie algebra over $\mathbb{R}$ and $\mathfrak{k}$ be a subalgebra of $\mathfrak{g}$. Let (j) be a collection of linear endomorphisms of $\mathfrak{g}$ and ω be a linear form on $\mathfrak{g}$. Then the system $\{\mathfrak{g}, \mathfrak{k}, (j), \omega\}$ is called a j-algebra if the following conditions are satisfied:

ji) $j \equiv j' \mod \mathfrak{k}$ for $j, j' \in (j)$.
$j\mathfrak{k} \subset \mathfrak{k}$ for $j \in (j)$.

jii) $j^2 \equiv -1 \mod \mathfrak{k}$.

jiii) $[k, jx] \equiv j[k, x] \mod \mathfrak{k}$, $k \in \mathfrak{k}$ $x \in \mathfrak{g}$.

jiv) $[jx, jy] \equiv j[jx, y] + j[x, jy] + [x, y] \mod \mathfrak{k}$, $x, y \in \mathfrak{g}$.

jv) $\omega([k, x]) = 0$ $k \in \mathfrak{k}$.

jvi) $\omega([jx, jy]) = \omega([x, y])$.

jvii) $\omega([jx, x]) > 0$, $x \notin \mathfrak{k}$.

Remark. ji) means that each j induces one and the same linear endomorphism on the vector space $\mathfrak{g}/\mathfrak{k}$. If jii) - jvii) are satisfied for a special j, then they are satisfied for all $j \in (j)$. The subalgebra $\mathfrak{k}$ may be (0).

Example. Let $\mathfrak{g} = \mathfrak{sl}(2, \mathbb{R}) = \{\begin{pmatrix} a & b \\ c & -a \end{pmatrix} ; a, b, c \in \mathbb{R}\}$. Consider the element $I = \begin{pmatrix} 0 & 1 \\ -1 & 0 \end{pmatrix} \in \mathfrak{g}$. Let $\mathfrak{k}$ be the set $\{x \in \mathfrak{g} : [I, x] = 0\}$. Then $\mathfrak{k}$ consists of all the matrices $\lambda I, \lambda \in \mathbb{R}$, and $\mathfrak{k}$ coincides with the Lie algebra $\mathfrak{o}(2)$ of the special orthogonal group $SO(2)$. Define a linear endomorphism j_o on $\mathfrak{sl}(2, \mathbb{R})$ by putting $j_o x = \frac{1}{2}[I, x]$. Then j_o satisfies $j\mathfrak{k} = 0$ and jii) - jiv). Put $\mathfrak{m} = \{x \in \mathfrak{g} : Ix = -xI\}$. Then we have $\mathfrak{g} = \mathfrak{k} + \mathfrak{m}$ (vector space direct sum). Let $x \in \mathfrak{g}$. Then x can be written in the form $x = x_1 + x_2 = \lambda(x_1)I + x_2$, where $x_1 = \lambda(x_1)I \in \mathfrak{k}$, $x_2 \in \mathfrak{m}$ and λ is the linear form on $\mathfrak{k}$. Put $\omega(x) = \lambda(x_1)$. Then ω satisfies jv)-jvii). We define (j) to be the set of all the linear endomorphisms j of $\mathfrak{g}$ such that $j \equiv j_o$ and $j\mathfrak{k} \subset \mathfrak{k}$. Then $\{\mathfrak{sl}(2, \mathbb{R}), \mathfrak{o}(2), (j), \omega\}$ is a j-algebra.

Definition 3.2. Let $\{\mathfrak{g}, \mathfrak{k}, (j), \omega\}$ and $\{\mathfrak{g}', \mathfrak{k}', (j'), \omega'\}$ be j-algebras. $\mathfrak{g}$ is called a j-subalgebra of $\mathfrak{g}'$ if the following conditions are satisfied:

i) $\mathfrak{g}$ is a subalgebra of $\mathfrak{g}'$.

ii) $\mathfrak{k} = \mathfrak{k}' \wedge \mathfrak{g}$.

iii) $jx \equiv j'x \mod \mathfrak{k}'$ $x \in \mathfrak{g}$, $j \in (j)$, $j' \in (j')$.

Furthermore if $\mathfrak{g}$ is an ideal of $\mathfrak{g}'$, then $\mathfrak{g}$ is called a <u>j-ideal</u> of $\mathfrak{g}'$.

<u>Definition 3.3.</u> Let $\{\mathfrak{g}, \mathfrak{k}, (j), \omega\}$ and $\{\mathfrak{g}', \mathfrak{k}', (j'), \omega'\}$ be j-algebras. A homomorphism f of $\mathfrak{g}$ into $\mathfrak{g}'$ is called a <u>j-homomorphism</u> if the following conditions are satisfied:

i) $f \circ j \equiv j' \circ f \mod \mathfrak{k}'$, $j \in (j)$, $j' \in (j')$.

ii) $f(\mathfrak{k}) = f(\mathfrak{g}) \wedge \mathfrak{k}'$.

In this case we see that $\{f(\mathfrak{g}), f(\mathfrak{k})\}$ becomes a j-subalgebra. If f is an isomorphism satisfying i), ii), then f is called a <u>j-isomorphism</u>. If f is an isomorphism of $\mathfrak{g}$ onto $\mathfrak{g}'$ satisfying i), ii), then $\mathfrak{g}$ is <u>j-isomorphic</u> to $\mathfrak{g}'$.

Let G/K be a complex homogeneous space, where G is a Lie group and K is a closed subgroup of G . Then there exists a collection (j) on the Lie algebra $\mathfrak{g}$ of G satisfying ji) - jiv). Furthermore we have

<u>Theorem 3.4.</u> (Koszul [11]). <u>Let</u> M=G/K <u>be a complex homogeneous space</u>. If M <u>is holomorphically equivalent to a bounded domain in</u> $\mathbb{C}^n$, <u>then there exists a linear form</u> ω <u>on the Lie algebra</u> $\mathfrak{g}$ <u>of</u> G <u>satisfying</u> jv) - jvii). <u>In particular</u>, $\{\mathfrak{g}, \mathfrak{k}, (j), \omega\}$ <u>is a j-algebra, where</u> $\mathfrak{k}$ <u>is the Lie algebra of</u> K .

<u>A sketch of the proof.</u> M has the Bergman metric g , which is a G-invariant Kähler metric. We denote by I the complex structure on M. And we define a 2-form ψ by putting $\psi(X,Y)=g(X,IY)$, where X,Y are vector fields on M . Let π be the canonical projection of G onto G/K . Then the pull-back $\pi^*\psi$ is a left-invariant 2-form on G . According to [11] there exists a G-invariant 1-form ω on G such that $\pi^*\psi=-d\omega$. ω can be considered as a linear form on $\mathfrak{g}$ and $\omega([x,y])= \pi^*\psi(x,y)$ holds for $x,y \in \mathfrak{g}$. Using this equality we can easily see that ω satisfies jv) - jvii).

<u>Remark.</u> In general, there are many j-algebras corresponding to one and the same homogeneous bounded domain D, since there are lots of Lie subgroups of $G_h(D)$ acting on D transitively.

Let $D(V,F) \subset R^c \times W$ be an affine homogeneous Siegel domain of type II. Let $\mathfrak{g}_a$ and $\mathfrak{t}$ be the Lie algebras of the Lie groups G_a and T, respectively. Now we want to consider the structures of $\mathfrak{g}_a$ and $\mathfrak{t}$. The Lie algebra of the group RW is canonically identified with the tangent space of RW at the identity $(0,0)$, which is identified with the direct sum $R+W$. Hence we will regard $R+W$ as the Lie algebra of RW. To distinguish these R,W from the subspaces R,W of $R^c \times W$, we denote the formers by $\underline{R}$ and $\underline{W}$. From the arguments in §1 and §2, we have

$$\left.\begin{aligned} \mathfrak{g}_a &= \mathfrak{h} + \underline{R} + \underline{W} = \mathfrak{k}_a + \mathfrak{t} \\ \mathfrak{h} &= \mathfrak{k}_a + \mathfrak{t}_1 \\ \mathfrak{t} &= \mathfrak{t}_1 + \underline{R} + \underline{W} \end{aligned}\right\} \text{(vector space direct sums)}$$

where $\mathfrak{h}$ and $\mathfrak{t}_1$ are the Lie algebras of H and T_1, respectively.

Let $g \in \mathfrak{g}_a$. We denote by D_g the vector field on $R^c \times W$ induced by the one-parameter group $\exp tg$. Hence the value of D_g at (z,u) is given by

$$D_g(z,u) = \lim_{t \to 0} \frac{1}{t}\big((\exp tg)(z,u) - (z,u)\big) \qquad \ldots(*)$$

<u>Proposition 3.5.</u>

(1) $D_h(x+iy,u) = \big(\rho_*(h)x + i\rho_*(h)y,\ \sigma_*(h)u\big) \qquad h \in \mathfrak{h}$

(2) $D_a(x+iy,u) = (a,0) \qquad a \in \underline{R}$

(3) $D_c(x+iy,u) = \big(2iF(u,c),c\big) \qquad c \in \underline{W}$

<u>where</u> ρ_*, σ_* <u>are representations of</u> $\mathfrak{h}$ <u>induced by</u> ρ,σ, <u>respectively,</u> <u>and</u> a,c <u>in the right hand sides are considered as elements in</u> R,W, <u>respectively, via the identification of the tangent space</u> $T_{(x+iy,u)}\big(D(V,F)\big)$ with $R^c \times W$.

<u>Proof.</u> (1) Since $G_a = H \cdot RW$, the one-parameter subgroup generated by h

is $(\exp th,0,0)$. By Theorem 1.13,4) we have

$$(\exp th,0,0)(x+iy,u) = \big(\rho(\exp th)x+i\rho(\exp th)y,\ \sigma(\exp th)u\big)$$

$$= \big(\exp t\rho_*(h)\cdot x+i\ \exp t\rho_*(h)\cdot y,\ \exp t\sigma_*(h)\cdot u\big)\ .$$

Using (*) and this equality we have (1).

(3) Using Theorem 1.13,5) it can be verified that $(1,0,tc)$ is the one-parameter subgroup generated by $c\varepsilon \underline{W}$. Theorem 1.13,4) shows that

$$(1,0,tc)(x+iy,u) = \big(x+iy+2tiF(u,c)+t^2 iF(c,c),\ u+tc\big)\ .$$

Hence by (*) we get (3).

(2) is easily seen.

<u>Proposition 3.6.</u> <u>The bracket relations in</u> $\mathfrak{g}_a$ <u>are given as follows:</u>

(1)	$[h,a] = \rho_*(h)a$		$h\varepsilon\mathfrak{k},\ a\varepsilon\underline{R}$
(2)	$[h,c] = \sigma_*(h)c$		$h\varepsilon\mathfrak{k},\ c\varepsilon\underline{W}$
(3)	$[c,c'] = -4\mathrm{Im}F(c',c)$		$c,c'\varepsilon\underline{W}$
(4)	$[a,c] = 0$		$a\varepsilon\underline{R}\quad c\varepsilon\underline{W}$
(5)	$[a,a'] = 0$		$a,a'\varepsilon\underline{R}$

<u>Proof.</u> To obtain these relations, we consider $\mathfrak{g}_a$ as a subalgebra of $\mathfrak{gl}(n+m+1,\mathbb{C})$, as is used in §2. Then we have

$$[x_1,x_2] = x_1x_2 - x_2x_1 \qquad (**)$$

for $x_1,x_2\varepsilon\ \mathfrak{g}_a$, where the multiplications in the right-hand side are those as matrices.

It follows from Proposition 3.5. that h,a and c can be represented as the following matrices in $\mathfrak{gl}(n+m+1,\mathbb{C})$:

$$h = \begin{pmatrix} \rho_*(h) & 0 & 0 \\ 0 & \sigma_*(h) & 0 \\ 0 & 0 & 0 \end{pmatrix}$$

$$a = \begin{pmatrix} 0 & 0 & a \\ 0 & 0 & 0 \\ 0 & 0 & 0 \end{pmatrix}$$

$$c = \begin{pmatrix} 0 & f(c) & 0 \\ 0 & 0 & c \\ 0 & 0 & 0 \end{pmatrix}$$

where $f(c)$ is the linear map $u \mapsto 2iF(u,c) \in R^c$, $u \in W$. Using these forms and (**), we get Proposition 3.6.

<u>Corollary 3.7.</u> $\rho_*(h) = \mathrm{ad}_{\underline{R}} h$, $\quad h \in \mathfrak{h}$

$\sigma_*(h) = \mathrm{ad}_{\underline{W}} h$, $\quad h \in \mathfrak{h}$

<u>Corollary 3.8.</u> <u>The following bracket relations are valid:</u>

$$[\mathfrak{h}, \underline{R}] \subset \underline{R} \qquad [\mathfrak{h}, \underline{W}] \subset \underline{W}$$

$$[\underline{W}, \underline{W}] \subset \underline{R} \qquad [\underline{R}, \underline{W}] = [\underline{R}, \underline{R}] = 0$$

<u>In particular</u> $\underline{R}$ <u>is an abelian ideal of</u> $\mathfrak{g}_a$.

These corollaries are immediate consequences of Proposition 3.6.

Using Proposition 3.5., we will identify $\underline{R}$ and $\underline{W}$ with the subspaces R and W of $R^c \times W$, respectively, in the following way. For each $X \in \mathfrak{g}_a$, put $\alpha(X) = X_e$, where X_e is the value of X at the identity $e \in G_a$. Then α is a linear isomorphism of $\mathfrak{g}_a$ onto the tangent space $T_e(G_a)$ of G_a at e. The Lie algebra $\mathfrak{k}_a$ is the isotropy subalgebra of $\mathfrak{g}_a$ at the point $z_o = (ir, 0) \in D(V,F)$. Let β be the natural projection of $T_e(G_a)$ onto $T_{z_o}(D(V,F))$, the tangent space at z_o. The identification of $\underline{R}$ (resp. $\underline{W}$) with R (resp. W) is given by the composite map $\pi = \beta \circ \alpha$. As is known in the theory of Lie transformation groups, we have

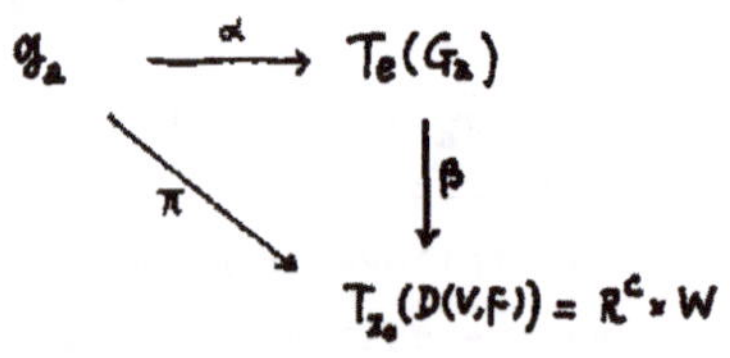

$$\pi(X) = D_X(z_o) \qquad X \in \mathfrak{g}_a .$$

Hence for $a \in \underline{R}$ and $c \in \underline{W}$ we see from Proposition 3.5. that

$$\pi(a) = D_a(ir,0) = (a,0) \in R$$

$$\pi(c) = D_c(ir,0) = (0,c) \in W .$$

Let J_{z_o} be the complex structure on $T_{z_o}(D(V,F))$. Then there exist linear endomorphisms j on $\mathfrak{g}_a$ such that

$$\begin{cases} D_{jg}(z_o) = J_{z_o} D_g(z_o) \qquad g \in \mathfrak{g}_a \\ j\mathfrak{k}_a \subset \mathfrak{k}_a \\ j\underline{W} \subset \underline{W} . \end{cases}$$

Let (j) be the collection of these j's. Then (j) satisfies ji) - jiv). The existence of the linear form ω satisfying jv) - jvii) is guaranteed by Theorem 3.4. Thus $\{\mathfrak{g}_a, \mathfrak{k}_a, (j), \omega\}$ is a j-algebra.

In the case of the Lie algebra $\mathfrak{t}$, π (restricted to $\mathfrak{t}$) ia a linear isomorphism of $\mathfrak{t}$ onto $T_{z_o}(D(V,F))$ and so there exists one and only one linear endomorphism j on $\mathfrak{t}$ such that

$$D_{jg}(z_o) = J_{z_o} D_g(z_o) , \qquad g \in \mathfrak{t}$$

and this satisfies

$$j\underline{W} \subset \underline{W} .$$

In fact, for each $c \in \underline{W}$, we have $D_{jc}(z_o) = J_{z_o} D_c(z_o) = i(0,c) = (0,ic) \in W$, which implies that $jc \in \pi^{-1}(W) = \underline{W}$. Thus the Lie algebra $\mathfrak{t}$, together with j and $\omega | \mathfrak{t}$, is a j-algebra.

<u>Definition 3.9.</u> Let D be a homogenueous bounded domain in $\mathbb{C}^n$, and D(V,F) be an affine homogeneous Siegel domain of type II holomorphically equivalent to D. Let T be the Iwasawa subgroup of $G_h(D)$ with respect to D(V,F). Then the Lie algebra $\mathfrak{t}$ with the j-algebra structure defined above is called the <u>Iwasawa j-algebra</u> of D (with respect to the realization D(V,F) of D).

<u>Remark.</u> We can prove that there exists a one-to-one correspondence between the set of all holomorphic equivalence classes of homogeneous bounded domains and the set of all j-isomorphic classes of Iwasawa j-algebras (cf. Kaneyuki [7]).

<u>Lemma 3.10.</u> 1) $\mathfrak{f} = \mathfrak{k}_a + j\underline{R}$ (vector space direct sum).

2) $\mathfrak{f}_1 = j\underline{R}$ <u>is valid in</u> $\mathfrak{f}$.

3) $[ja,r] = a$ <u>for all</u> $a \in \underline{R}$.

<u>Proof.</u> 1) First we will prove $j\underline{R} \subset \mathfrak{f}$. Take $a \in \underline{R}$. Then

$$D_{ja}(z_o) = J_{z_o} D_a(z_o) = i(a,0) = (ia,0)$$

where $z_o=(ir,0)$. On the other hand the group $\rho(H)$ acts on V transitively as linear transformations. And the value at $r \in V$ of the vector field on V induced by $h \in \mathfrak{f}$ is $\rho_*(h)r$. Since $\rho(H)$ is transitive on V, $\{\rho_*(h)r : h \in \mathfrak{f}\}$ spans R . Hence there exists $h \in \mathfrak{f}$ such that $\rho_*(h)r=a$, from which it follows that

$$D_{h-ja}(z_o) = D_h(z_o) - D_{ja}(z_o) = (i\rho_*(h)r,0) - (ia,0) = 0$$

Hence $h-ja \in \mathfrak{k}_a$, which implies $ja \in \mathfrak{f}$. Furthermore $\mathfrak{k}_a \cap j\underline{R}=0$ holds. In fact, take $ja \in \mathfrak{k}_a \cap j\underline{R}$, where $a \in \underline{R}$. Then $0=D_{ja}(z_o)=(ia,0)$. So $a=0$. Hence the subspace $\mathfrak{k}_a+j\underline{R}$ of $\mathfrak{f}$ is the vector space direct sum of $\mathfrak{k}_a$ and $j\underline{R}$. And we see easily that j is one-to-one on $\underline{R}$. So we have $\dim(\mathfrak{k}_a+j\underline{R})=\dim\mathfrak{k}_a+\dim\underline{R}=\dim\mathfrak{f}$, which shows (1).

2) is analogously proved.

3) $(ia,0)=J_{z_o}D_a(z_o)=D_{ja}(z_o)=D_{ja}(ir,0)=(i\rho_*(ja)r,0)$
$=(i[ja,r],0)$ (cf. Corollary 3.7.).

QED

Thus we have proved the following two theorems:

<u>Theorem 3.11.</u> <u>Let</u> $D(V,F) \subset R^c \times W$ <u>be an affine homogeneous Siegel domain of type II</u>. <u>Then the Iwasawa j-algebra</u> $\mathfrak{f}$ <u>of</u> $D(V,F)$ <u>can be decomposed into the vector space direct sum</u>:

$$\mathfrak{f} = j\underline{R} + \underline{R} + \underline{W} \quad ,$$

<u>where</u> $\underline{R}$ <u>is an abelian ideal and</u>

$$[j\underline{R},\underline{W}] \subset \underline{W} \qquad [\underline{W},\underline{W}] \subset \underline{R} \qquad j\underline{W} = \underline{W}$$

$$[\underline{R},\underline{W}] = 0 \qquad [j\underline{R},j\underline{R}] \subset j\underline{R} \quad .$$

Moreover there exists $r \in \underline{R}$ *such that*

$$[ja, r] = a \qquad \ldots\ldots(*)$$

for all $a \in \underline{R}$.

Theorem 3.12 ([20]) . *Under the same assumption as in Theorem 3.11, the Lie algebra* $\mathfrak{g}_a$ *of the affine automorphism group* G_a *has the following structure:*

$$\mathfrak{g}_a = \mathfrak{k}_a + j\underline{R} + \underline{R} + \underline{W} , \qquad \text{(vector space direct sum)}$$

where $\underline{R}$ *is an abelian ideal and*

$$[\mathfrak{k}_a + j\underline{R}, \underline{W}] \subset \underline{W} ,$$

$$[\underline{W}, \underline{W}] \subset \underline{R} , \quad [\underline{R}, \underline{W}] = 0 .$$

Moreover there exists $r \in \underline{R}$ *satisfying* (*) *in Theorem 3.11.*

Proposition 3.13. *Under the identifications of* R *with* $\underline{R}$ *and* W *with* $\underline{W}$ *, we have*

1) $V = (Ad_{\underline{R}} T) \cdot r$,
where T *is the simply connected Lie group generated by* $\mathfrak{t}$.

2) $F(u,v) = \frac{1}{4}([ju,v] + i[u,v]) \in \underline{R} + i\underline{R}$.

3) *The domain* $D(V,F)$ *is re-constructed in* $\underline{R} + i\underline{R} + W$ *as follows:*

$$D(V,F) = \{x+iy+u \in \underline{R}+i\underline{R}+W \ ; \ y - \tfrac{1}{4}[ju,u] \in V\} .$$

Proof. 1) Since $z_o = (ir, 0)$ is in $D(V,F)$, we have $r \in V$. By Corollary 3.7 and Theorem 3.11 we have $ad_{\underline{R}} \mathfrak{t} = ad_{\underline{R}} \mathfrak{t}_1 = \rho_*(\mathfrak{t}_1)$, which implies $Ad_{\underline{R}} T = \rho(T_1)$. Moreover $\rho(T_1)$ acts on V transitively.

2) We have the following relations:

$$[u,v] = -4\mathrm{Im}F(v,u) = 4\mathrm{Im}F(u,v)$$

$$[ju,v] = -4\mathrm{Im}F(v,iu) = 4\mathrm{Im}\,iF(v,u) = 4\mathrm{Re}F(u,v) .$$

QED

In the following each section we will always identify $\underline{R}$ and $\underline{W}$ with R and W , respectively.

§4. Universal j-algebras

Let X be a 2m-dimensional vector space over $\mathbb{R}$ and I be a linear endomorphism of X, and ρ be an alternating bilinear form on X. Then the triple (X,I,ρ) is called a symplectic vector space if the following conditions are satisfied:

si) $I^2 = -1$

sii) $\rho(Ix,Iy) = \rho(x,y)$

siii) $\rho(Ix,x)>0$ for $x\neq 0$.

For a symplectic vector space (X,I,ρ), a linear endomorphism p of X is called symplectic if p satisfies

$$\rho(pu,v) + \rho(u,pv) = 0$$

for all $u,v\in X$. We denote by $\mathfrak{sp}(X)$ the set of all symplectic endomorphisms, which is a subalgebra of $\mathfrak{gl}(X)$. Since $I\in \mathfrak{sp}(X)$, we can define the linear endomorphism j on $\mathfrak{sp}(X)$ by

$$j(p) = \frac{1}{2}[I,p] \quad , \qquad p\in \mathfrak{sp}(X) .$$

We can easily see that $\mathfrak{sp}(X)$ is isomorphic to $\mathfrak{sp}(m,\mathbb{R})$, the Lie algebra of the symplectic group $Sp(m,\mathbb{R})$. Let us put

$$\mathfrak{u}(X) = \{p\in \mathfrak{sp}(X) \ ; \ Ip = pI\}$$

$$\mathfrak{p}(X) = \{p\in \mathfrak{sp}(X) \ ; \ Ip+pI = 0\} .$$

Then we have

$$\left.\begin{array}{l} [\mathfrak{u}(X), \mathfrak{u}(X)] \subset \mathfrak{u}(X) \\ [\mathfrak{u}(X), \mathfrak{p}(X)] \subset \mathfrak{p}(X) \\ [\mathfrak{p}(X), \mathfrak{p}(X)] \subset \mathfrak{u}(X) \end{array}\right\} \qquad (*)$$

Also we have the decomposition

$$\mathfrak{sp}(X) = \mathfrak{u}(X) + \mathfrak{p}(X) .$$

It can be proved (cf. Lemma 8.1 and 8.2) that this decomposition is a Cartan decomposition of $\mathfrak{sp}(X)$ and that $\mathfrak{u}(X)$ is a maximal compact subalgebra. Since $\mathfrak{sp}(X) \cong \mathfrak{sp}(m,\mathbb{R})$, the center of $\mathfrak{u}(X)$ is one-dimensional and is spanned by I. Hence we get

$$\mathfrak{sp}(X) = \{\lambda I\} + \mathfrak{u}(X)' + \mathfrak{p}(X)$$

where $\lambda\varepsilon\mathbb{R}$ and $\mathfrak{u}(X)'$ is the commutator subalgebra of $\mathfrak{u}(X)$. Each $x\varepsilon\mathfrak{g}(X)$ can be written uniquely in the form $x=\lambda I+a+b$, $a\varepsilon\mathfrak{u}(X)'$, $b\varepsilon\mathfrak{p}(X)$. We define the linear form ω_o on $\mathfrak{g}(X)$ to be

$$\omega_o(x) = \lambda .$$

Then we have

<u>Proposition 4.1.</u> ([15]). $\{\mathfrak{g}(X), \mathfrak{u}(X), j, \omega_o\}$ <u>is a j-algebra</u>.

<u>Proof.</u> From the definition of j it follows that j vanishes on $\mathfrak{u}(X)$ and that j leaves $\mathfrak{p}(X)$ stable. The properties (ji)-(jiv) of a j-algebra can be easily verified by using (*).

jv) Let $k=\lambda I+a\varepsilon\mathfrak{u}(X)$ and $x=\lambda' I+a'+b'\varepsilon\mathfrak{g}(X)$ where $a,a'\varepsilon\mathfrak{u}(X)'$ and $b'\varepsilon\mathfrak{p}(X)$. Then $[k,x] = \lambda[I,b'] + [a,b'] + [a,a']$. Hence by (*) we get $\omega_o([k,x])=0$.

jvi) Let $x=\lambda I+a+b$, $x'=\lambda' I+a'+b'\varepsilon\mathfrak{g}(X)$, where $a, a'\varepsilon\mathfrak{u}(X)'$ and $b, b'\varepsilon\mathfrak{p}(X)$. Then

$$[jx,jx'] = \frac{1}{4}[[I,b],[I,b']] = [Ib,Ib'] = IbIb'-Ib'Ib = [b,b'] .$$

Hence by jv) we have

$$\omega_o([jx,jx']) = \omega_o([b,b']) = \omega_o([x,x']) .$$

jvii) Let $x=\lambda I+a+b$, $a\varepsilon\mathfrak{u}(X)'$, $b\varepsilon\mathfrak{p}(X)$. Then

$$[jx,x] = \frac{1}{2}([[I,b], \lambda I+a] + [[I,b],b]) .$$

Hence

$$\omega_o([jx,x]) = \frac{1}{2}\omega_o([[I,b],b]) .$$

We can write $[[I,b],b]$ as $[[I,b],b]=\mu I+c$, $c\varepsilon\mathfrak{u}(X)'$. To verify jvii) it suffices to show that if $b\neq 0$, then $\mu>0$. Suppose $b\neq 0$. Then $b'=[I,b]\neq 0$. Let B be the Killing form of $\mathfrak{g}(X)$. Then

$$B([[I,b],b],I) = -B([I,b],[I,b]) = -B(b',b') ;$$

the left-hand side is equal to $\mu B(I,I)+B(c,I)$. Since $B(\lambda I, \mathfrak{u}(X)')=0$ for $\lambda\varepsilon\mathbb{R}$, we have $B(c,I)=0$. So we have

$$\mu B(I,I) = -B(b',b') .$$

Moreover, since B is negative definite on $\mathfrak{u}(X)$ and positive definite on $\mathfrak{f}(X)$, we conclude $\mu>0$. Thus we proved

$$\omega_0([jx,x]) = \frac{1}{2}\mu>0 .$$

QED

Definition 4.2. ([20]) . A j-algebra $\{\mathfrak{g},\mathfrak{k},(j),\omega\}$ is called proper if any compact semi-simple j-subalgebra $\mathfrak{g}_0$, more precisely $\{\mathfrak{g}_0,\mathfrak{k}_0,(j'),\omega'\}$, is contained in $\mathfrak{k}$.

Lemma 4.3. ([20]) . Let $\{\mathfrak{g},\mathfrak{k},(j),\omega\}$ be a j-algebra of a homogeneous bounded domain D (cf. Theorem 3.4.) . Then it is proper .

Proof. D can be represented as a complex coset space G/K , where G is a connected Lie group corresponding to $\mathfrak{g}$ and K is a Lie subgroup of G whose Lie algebra is $\mathfrak{k}$. Let $\mathfrak{g}_0$ be a compact semi-simple j-subalgebra of $\mathfrak{g}$ and G_0 be the analytic subgroup of G generated by $\mathfrak{g}_0$. Then G_0 is compact and semi-simple. The orbit $G_0\cdot o$, o being the origin of G/K , is compact, and is also a complex submanifold of D , since $\mathfrak{g}_0$ is a j-subalgebra. Since D is a domain of holomorphy by a theorem of Thullen, any compact analytic set in D is zero-dimensional, which implies $G_0\cdot o=o$. Thus we get $\mathfrak{g}_0\subset\mathfrak{k}$.

QED

Let D_1 be a homogeneous bounded domain in $\mathbb{C}^n$. Suppose that D_1 is realized as the Siegel domain D(V,F) of type II (not of type I) in $R^c\times W$. Let $\{\mathfrak{f}_1,j,\omega_1\}$ be its Iwasawa j-algebra, and

$$\mathfrak{f}_1 = jR + R + W$$

be the decomposition given in Theorem 3.11. Then $W\neq(0)$. We denote by j_W the restriction of j to W . For each $u,v\in W$ we put $\rho(u,v) =\omega_1([u,v])$. Then (W,j_W,ρ) is a symplectic vector space, and by Proposition 4.1. $\{\mathfrak{g}(W),\mathfrak{u}(W),j,\omega_0\}$ is a j-algebra. Let $\tilde{\mathfrak{r}}_1$ be the subalgebra of $\mathfrak{g}(W)$ consisting of all the elements p satisfying the conditions

$$[pu,v] + [u,pv] = 0 , \qquad u,v\in W$$
$$[p,ad_W jR] = 0 .$$

From jiv) and the bracket relations in Theorem 3.11 it follows that $j_W \in \tilde{\mathfrak{r}}_1$. In particulare $\tilde{\mathfrak{r}}_1$ is a non-zero subalgebra of $\mathfrak{sp}(W)$.

<u>Lemma 4.4. The Lie algebra $\tilde{\mathfrak{r}}_1$ is reductive.</u>

<u>Proof.</u> We can easily see that $\operatorname{Ker} j = \mathfrak{u}(W)$ and $\operatorname{Im} j = \mathfrak{p}(W)$. Since $j_W \in \tilde{\mathfrak{r}}_1$, $\tilde{\mathfrak{r}}_1$ is stable under j . So we get

$$\tilde{\mathfrak{r}}_1 = \tilde{\mathfrak{r}}_1 \cap \mathfrak{u}(W) + \tilde{\mathfrak{r}}_1 \cap \mathfrak{p}(W) .$$

Let σ be the Cartan involution of $\mathfrak{sp}(W)$ with respect to the Cartan decomposition $\mathfrak{sp}(W) = \mathfrak{u}(W) + \mathfrak{p}(W)$. Then, from the decomposition above it follows that $\tilde{\mathfrak{r}}_1$ is stable under σ . Hence by a theorem of Borel, Harish-Chandra [2] , $\tilde{\mathfrak{r}}_1$ is reductive.

QED

We put

$$\tilde{\mathfrak{k}}_1 = \tilde{\mathfrak{r}}_1 \cap \mathfrak{u}(W)$$

$$\mathfrak{p} = \tilde{\mathfrak{r}}_1 \cap \mathfrak{p}(W) .$$

For a while we will assume $\mathfrak{p} \neq (0)$. This condition is satisfied for a large class of $\mathfrak{r}_1$ including important examples.

<u>Lemma 4.5. The pair $\{\tilde{\mathfrak{r}}_1, \tilde{\mathfrak{k}}_1\}$ is a j-subalgebra of</u> $\{\mathfrak{sp}(W), \mathfrak{u}(W), j, \omega_0\}$; <u>the center $\mathfrak{z}$ of $\tilde{\mathfrak{r}}_1$ is contained in $\tilde{\mathfrak{k}}_1$</u> .

<u>Proof.</u> Since $\tilde{\mathfrak{r}}_1$ is j-stable and $\tilde{\mathfrak{k}}_1 = \mathfrak{u}(W) \cap \tilde{\mathfrak{r}}_1$, we get the first assertion. Now let $z \in \mathfrak{z}$. Then $[jz, z] = 0$.
This implies $\omega_0([jz,z]) = 0$. By jviii) we have $z \in \tilde{\mathfrak{k}}_1$.

QED

Since $\tilde{\mathfrak{r}}_1$ is reductive, we have the decomposition

$$\tilde{\mathfrak{r}}_1 = \mathfrak{z} + \tilde{\mathfrak{r}} ,$$

$\tilde{\mathfrak{r}}$ being the commutator subalgebra of $\tilde{\mathfrak{r}}_1$ which is semi-simple. Moreover

$$\tilde{\mathfrak{k}}_1 = \mathfrak{z} + \tilde{\mathfrak{k}}_1 \cap \tilde{\mathfrak{r}} .$$

Here we put $\tilde{\mathfrak{k}} = \tilde{\mathfrak{k}}_1 \cap \tilde{\mathfrak{r}}$. Then we have

<u>Lemma 4.6.</u> $\{\tilde{\mathfrak{s}}, \tilde{\mathfrak{k}}\}$ <u>is a j-subalgebra of</u> $\{\tilde{\mathfrak{s}}_1, \tilde{\mathfrak{k}}_1, j, \omega_0\}$; <u>and we have the following Cartan decomposition</u>

$$\tilde{\mathfrak{s}} = \tilde{\mathfrak{k}} + \mathfrak{p} \ .$$

<u>Proof.</u> Take $x \in \tilde{\mathfrak{s}}$. Then

$$jx = \frac{1}{2}[j_W, x] \in [\tilde{\mathfrak{s}}_1, \tilde{\mathfrak{s}}_1] = \tilde{\mathfrak{s}} \ ,$$

which implies that $\tilde{\mathfrak{s}}$ is j-stable. Hence we get the first assertion. From a lemma of Borel, Harish-Chandra [2] it follows that the restriction of the Cartan involution σ to $\tilde{\mathfrak{s}}$ is again a Cartan involution of $\tilde{\mathfrak{s}}$. So we get the Cartan decomposition by σ :

$$\tilde{\mathfrak{s}} = \tilde{\mathfrak{s}} \cap \tilde{\mathfrak{k}}_1 + \tilde{\mathfrak{s}} \cap \mathfrak{p} = \tilde{\mathfrak{k}} + \tilde{\mathfrak{s}} \cap \mathfrak{p} \ .$$

Moreover $\dim(\tilde{\mathfrak{s}} \cap \mathfrak{p}) = \dim \tilde{\mathfrak{s}} - \dim \tilde{\mathfrak{k}} = \dim \tilde{\mathfrak{s}}_1 - \dim \tilde{\mathfrak{k}}_1 = \dim \mathfrak{p}$, which proves the second assertion.

QED

Let $\mathfrak{k}_0$ be a maximal (semi-simple) ideal of $\tilde{\mathfrak{s}}$ contained in $\tilde{\mathfrak{k}}$. Then

$$\tilde{\mathfrak{s}} = \mathfrak{k}_0 + \mathfrak{k} + \mathfrak{p} \ ,$$

where $\mathfrak{k} = [\mathfrak{p}, \mathfrak{p}]$. Put $\mathfrak{s} = \mathfrak{k} + \mathfrak{p}$. Then $\mathfrak{s}$ is a semi-simple ideal of non-compact type of $\tilde{\mathfrak{s}}$ and $\mathfrak{k}$ is a maximal compact subalgebra of $\mathfrak{s}$. $\{\mathfrak{s}, \mathfrak{k}\}$ is an effective j-subalgebra of $\{\tilde{\mathfrak{s}}, \tilde{\mathfrak{k}}\}$.

We can take the following linear form as ω_1

$$\mathfrak{t}_1 \ni x \mapsto \mathrm{Tr}\, \mathrm{ad}_R jx \ .$$

In fact, using the roots theory of a j-algebra (cf. [20]) it can be verified that the linear form satisfies the properties jvi) and jvii). Hence the bracket relations in Theorem 3.11 show that ω_1 vanishes on W .

Let us consider the vector space direct sums

$$\tilde{\mathfrak{g}}_1 = \mathfrak{t}_1 + \tilde{\mathfrak{s}}_1$$

$$\mathfrak{g} = \mathfrak{t}_1 + \mathfrak{s} \ .$$

We define the bracket relations in $\tilde{\mathfrak{g}}_1$ and $\mathfrak{g}$ as follows:

$$[\tilde{\mathfrak{s}}_1, R + jR] = 0 \ ,$$

$$[p,u] = pu \ , \quad p \in \tilde{\mathfrak{r}}_1 \ , \quad u \in W \ .$$

The direct computations show that $\tilde{\mathfrak{g}}_1$ is a Lie algebra with respect to the above bracket relations. Obviously $\mathfrak{g}$ is a subalgebra of $\tilde{\mathfrak{g}}_1$, and $\mathfrak{t}_1$ is an ideal of $\tilde{\mathfrak{g}}_1$. We define the linear endomorphism j on $\tilde{\mathfrak{g}}_1$ to be the direct sum of j on $\mathfrak{t}_1$ and j on $\tilde{\mathfrak{r}}_1$, and also define the linear form ω on $\tilde{\mathfrak{g}}_1$ to be the direct sum of ω_1 and ω_o. Then we have

<u>Proposition 4.7.</u> ([15]) . $\{\tilde{\mathfrak{g}}_1, \tilde{\mathfrak{k}}_1, j, \omega\}$ <u>is an effective j-algebra</u>; $\{\mathfrak{g}, \mathfrak{k}, j, \omega\}$ <u>is an effective j-subalgebra</u>. <u>Moreover</u> $\mathfrak{t}_1$ <u>is a j-ideal of</u> $\tilde{\mathfrak{g}}_1$; $\{\tilde{\mathfrak{r}}_1, \tilde{\mathfrak{k}}_1\}$ <u>and</u> $\{\mathfrak{r}, \mathfrak{k}\}$ <u>are j-subalgebras of</u> $\tilde{\mathfrak{g}}_1$ <u>and</u> $\mathfrak{g}$, <u>respectively</u>.

<u>Proof.</u> Using the fact that ω_1 vanishes on W, we can prove by direct computations that $\{\tilde{\mathfrak{g}}_1, \tilde{\mathfrak{k}}_1, j, \omega\}$ is a j-algebra. To prove the effectivity let $\mathfrak{a}$ be an ideal of $\tilde{\mathfrak{g}}_1$ contained in $\tilde{\mathfrak{k}}_1$. Then we have $[\mathfrak{a}, \mathfrak{t}_1] \subset \mathfrak{a} \cap \mathfrak{t}_1 = (0)$. Take $p \in \mathfrak{a}$. Then $pu = [p,u] = 0$ for all $u \in W$. Hence $p=0$, which implies $\mathfrak{a} = (0)$. Other assertions are trivial.

QED

<u>Definitions 4.8.</u> The j-algebra $\{\mathfrak{g}, \mathfrak{k}, j, \omega\}$ is called the <u>universal j-algebra</u> of $\mathfrak{t}_1$; the j-subalgebra $\{\mathfrak{r}, \mathfrak{k}\}$ is called the <u>classifying j-algebra</u> of $\mathfrak{t}_1$.

<u>Remark.</u> If the homogeneous bounded domain D_1 is realized as a Siegel domain of type I, or if $\mathfrak{v} = (0)$, then we have $\mathfrak{r} = (0)$ and consequently $\mathfrak{g} = \mathfrak{t}_1$.

<u>Definitions 4.9.</u> Let $\mathfrak{g}$ be a Lie algebra over $\mathbb{R}$, j a linear endomorphism of $\mathfrak{g}$, and ω be a linear form on $\mathfrak{g}$. The triple $\{\mathfrak{g}, j, \omega\}$ is called a <u>normal j-algebra</u> if the following conditions are satisfied:

Ni) $\mathfrak{g}$ is solvable and $\mathrm{ad}\,\mathfrak{g}$ is $\mathbb{R}$-triangular.

Nii) $j^2 = -1$

Niii) $[jx,jy] = j[jx,y] + j[x,jy] + [x,y]$

Viv) $\omega([jx,jy]) = \omega([x,y])$

Nv) $\omega([jx,x]) > 0$, $x \neq 0$.

As it is obvious from the above definition and Proposition 2.8, the Iwasawa j-algebra of a homogeneous bounded domain is normal. It is known in Piatetski-Sapiro [15] that a normal j-algebra is the Iwasawa j-algebra of a certain homogeneous bounded domain.

Let $\{\mathfrak{f}_o, j, \omega\}$ be the Iwasawa j-algebra of a homogeneous bounded domain, and let $\mathfrak{f}_1$ be a j-ideal of $\mathfrak{f}_o$. We define the hermitian inner product h by

$$h(x,y) = \omega([jx,y]) + i\omega([x,y])$$

for $x,y \in \mathfrak{f}_o$. Let $\mathfrak{f}_2$ be the orthogonal complement of $\mathfrak{f}_1$ with respect to h. As is known in [15], $\mathfrak{f}_2$ is a j-subalgebra and we have the semi-direct sum

$$\mathfrak{f}_o = \mathfrak{f}_1 + \mathfrak{f}_2 . \qquad (\#)$$

Furthermore $\mathfrak{f}_1$ and $\mathfrak{f}_2$ are normal. Since $\mathfrak{f}_1$ is the Iwasawa j-algebra of a homogeneous bounded domain, by Theorem 3.11 $\mathfrak{f}_1$ can be written as

$$\mathfrak{f}_1 = jR + R + W \quad ;$$

the subspaces in the right-hand side satisfy the bracket relations in Theorem 3.11. On the other hand, applying Theorem A2 and A3 in the Appendix to $\mathfrak{f}_1$, we get the decomposition

$$\mathfrak{f}_1 = \mathfrak{f}_1^o + \mathfrak{f}_1^1 + \mathfrak{f}_1^{\frac{1}{2}}$$

as a graded Lie algebra. By Proposition 5.1 in [7] we get

$$\mathfrak{f}_1^o = jR , \quad \mathfrak{f}_1^1 = R , \quad \mathfrak{f}_1^{\frac{1}{2}} = W .$$

Furthermore, by a result of Piatetski-Sapiro [15] we have

$$\begin{aligned} [\mathfrak{f}_2, jR+R] &= 0 \\ [\mathfrak{f}_2, W] &\subset W . \end{aligned} \qquad (\natural)$$

Since $\mathfrak{f}_o$ is normal, $ad_W \mathfrak{f}_2$ is $\mathbb{R}$-triangular.

<u>Lemma 4.12.</u> <u>A maximal $\mathbb{R}$-triangular subalgebra $\mathfrak{f}^*$ of $\tilde{\mathfrak{g}}_1$ is contained in $\mathfrak{f}$.</u>

<u>Proof.</u> We have

$$\tilde{\mathfrak{s}}_1 = \mathfrak{z} + \mathfrak{k}_0 + \mathfrak{s} \qquad \text{(direct sum)} \qquad (\#\#)$$

$$\tilde{\mathfrak{k}}_1 = \mathfrak{z} + \mathfrak{k}_0 + \mathfrak{k} \ . \qquad \text{(direct sum)}$$

Let $\mathfrak{t}_s$ be a maximal R-triangular subalgebra of $\mathfrak{s}$. Then $\mathfrak{s} = \mathfrak{k} + \mathfrak{t}_s$ and so $\tilde{\mathfrak{s}}_1 = \tilde{\mathfrak{k}}_1 + \mathfrak{t}_s$. Since $\tilde{\mathfrak{k}}_1 = \tilde{\mathfrak{s}}_1 \cap \mathfrak{u}(W)$, $\tilde{\mathfrak{k}}_1$ is a compact subalgebra of $\tilde{\mathfrak{s}}_1$. Hence the semi-direct sum $\tilde{\mathfrak{s}}_1 = \tilde{\mathfrak{k}}_1 + \mathfrak{t}_s$ implies that $\mathfrak{t}_s$ is a maximal R-triangular subalgebra of $\tilde{\mathfrak{s}}_1$. By the conjugacy of maximal R-triangular subalgebras (cf. Vinberg [19]) there exists an element g of the adjoint group of the Lie group generated by $\tilde{\mathfrak{s}}_1$ such that $g\mathfrak{t}_s = \mathfrak{t}^*$. To prove $\mathfrak{t}^* \subset \mathfrak{s}$, it suffices to show that

$$\mathrm{Ad}\exp(X_1+X_2+X_3)\mathfrak{t}_s \subset \mathfrak{s}$$

for $X_1 \in \mathfrak{z}$, $X_2 \in \mathfrak{k}_0$, $X_3 \in \mathfrak{s}$. For this let $X \in \mathfrak{t}_s$. Then by (##) we have

$$\begin{aligned}\mathrm{Ad}\exp(X_1+X_2+X_3)X &= \mathrm{Ad}\exp X_3 \cdot \mathrm{Ad}\exp X_2 \cdot \mathrm{Ad}\exp X_1 \cdot X \\ &= (\mathrm{Ad}\exp X_3)X \in \mathfrak{s}\ .\end{aligned}$$

QED

Thus we have a theorem of Piatetski-Sapiro which is stronger than the original one in [15] ;

Theorem 4.13. Let $\mathfrak{g}_0$ be an Iwasawa j-algebra and $\mathfrak{g}_1$ a j-ideal of $\mathfrak{g}_0$ and $\mathfrak{g}_0 = \mathfrak{g}_1 + \mathfrak{g}_2$ be the decomposition in (#) . Let $\mathfrak{g} = \mathfrak{g}_1 + \mathfrak{s}$ be the universal j-algebra of $\mathfrak{g}_1$. Then there exists a natural j-homomorphism λ of $\mathfrak{g}_2$ into $\mathfrak{s}$ such that

$$[\lambda(t), x] = [t, x]$$

for $t \in \mathfrak{g}_2$, $x \in \mathfrak{g}_1$.

Proof. We must recall the decomposition $\mathfrak{g}_1 = jR + R + W$ given before. Let us define λ as

$$\lambda(t) = \mathrm{ad}_W t\ , \quad t \in \mathfrak{g}_2\ .$$

Then we have, for $t \in \mathfrak{g}_2$, $u, v \in W$,

$$[\lambda(t)u, v] + [u, \lambda(t)v] = [[t,u], v] + [u, [t,v]] = [t, [u,v]] = 0.$$

For $a \in R$, we have (cf. (4))

$$(\mathrm{ad}_W ja)\lambda(t)u = [ja,[t,u]] = [[ja,t],u] + [t,[ja,u]]$$
$$= \lambda(t)(\mathrm{ad}_W ja)u \ .$$

Hence $\lambda(\mathfrak{t}_2) \subset \tilde{\mathfrak{r}}_1$. On the other hand $\lambda(\mathfrak{t}_2)$ is R-triangular. So it is contained in a maximal R-triangular subalgebra of $\tilde{\mathfrak{r}}_1$. From Lemma 4.12 it follows that $\lambda(\mathfrak{t}_2) \subset \mathfrak{r}$. We will show that λ is a j-homomorphism. We must show $j\lambda(t) \equiv \lambda(jt) \bmod \mathfrak{k}$ for $t \in \mathfrak{t}_2$. Since the both-hand sides belong to $\mathfrak{r}$, it is enough to verify $j\lambda(t) \equiv \lambda(jt) \bmod \mathfrak{u}(W)$, or equivalently

$$\left[\lambda(jt) - \frac{1}{2}[j_W,\lambda(t)] \ ,j_W\right] = 0 \ .$$

This is equivalent to the condition jiv) in $\mathfrak{t}_0$ which is always true. To show $[\lambda(t),x] = [t,x]$ for $t \in \mathfrak{t}_2$, $x \in \mathfrak{t}_1$, let us write x in the form $x=a+jb+u$, $a,b \in R$, $u \in W$. So we get

$$[\lambda(t),x] = [\lambda(t),u] = \lambda(t)u = [t,u] = [t,x] \ .$$

QED

<u>Definition.</u> The j-homomorphism λ is called the <u>classifying j-homomorphism</u>.

§5. Universal Domains

Let $\mathfrak{t}_1$ be the Iwasawa j-algebra of a homogeneous bounded domain D_1, $\{\mathfrak{g}, \mathfrak{k}, j, \omega\}$ be the universal j-algebra of $\mathfrak{t}_1$. Then $\mathfrak{g}$ can be written as $\mathfrak{g} = \mathfrak{t}_1 + \mathfrak{s}$ (cf. §4). Let $\mathfrak{t}_s$ be a maximal $\mathbb{R}$-triangular subalgebra of $\mathfrak{s}$. We have $\mathfrak{s} = \mathfrak{k} + \mathfrak{t}_s$ and it follows that $\mathfrak{t}_s$ is a normal j-subalgebra of $\mathfrak{s}$.

Lemma 5.1. The subspace $\mathfrak{t} = \mathfrak{t}_1 + \mathfrak{t}_s$ of $\mathfrak{g}$ is a normal j-subalgebra of $\mathfrak{g}$.

Proof. It is clear that $\mathfrak{t}$ is a j-subalgebra of $\mathfrak{g}$. Since $\mathfrak{t}_1$ and $\mathfrak{t}_s$ are solvable and $\mathfrak{t}_1$ is an ideal, $\mathfrak{t}$ is also solvable. To prove that $\mathrm{ad}\,\mathfrak{t}$ is $\mathbb{R}$-triangular, it suffices to show that each eigenvalue of $\mathrm{ad}\, t$ is real for every $t \in \mathfrak{t}$. t can be written in the form $t = t_1 + t_2$, $t_1 \in \mathfrak{t}_1$, $t_2 \in \mathfrak{t}_s$. Let $\mathfrak{t}_1 = jR + R + W$ be the decomposition in Theorem 3.11. Then we can write t_1 as $t_1 = ja + b + u$, where $a, b \in R$, $u \in W$. $\mathrm{ad}\, t$ can be represented as the matrix

$$\begin{pmatrix} \mathrm{ad}_R ja & * & * & 0 \\ 0 & \mathrm{ad}_W t_2 + \mathrm{ad}_W ja & * & * \\ 0 & 0 & \mathrm{ad}_{jR} ja & 0 \\ 0 & 0 & 0 & \mathrm{ad}\, t_2 \end{pmatrix} .$$

Since $\mathrm{ad}\,\mathfrak{t}_1$ and $\mathrm{ad}\,\mathfrak{t}_s$ are $\mathbb{R}$-triangular, all eigenvalues of $\mathrm{ad}_R ja$, $\mathrm{ad}_W ja$, $\mathrm{ad}_{jR} ja$ and $\mathrm{ad}\, t_2$ are real. On the other hand we have $(\mathrm{ad}_W t_2)u = [t_2, u] = t_2 u$ for $u \in W$ and $[\mathrm{ad}_W t_2, \mathrm{ad}_W ja] = 0$ holds. Hence it follows that each eigenvalue of $\mathrm{ad}_W t_2 + \mathrm{ad}_W ja$ is real.

QED

From what is mentioned in §4, to the normal j-algebra $\mathfrak{t}$ there corresponds the homogeneous bounded domain D such that $\mathfrak{t}$ is its Iwasawa j-algebra.

Definition 5.2. The homogeneous bounded domain D is called the universal domain of the homogeneous bounded domain D_1.

Examples. 1) Let $V = H^+(2, \mathbb{R})$ and $W = \mathbb{C}$. We define a $H^+(2, \mathbb{R})$-hermitian form F as

$$F(u,v) = \begin{pmatrix} u\bar{v} & 0 \\ 0 & 0 \end{pmatrix}$$

for $u,v \in \mathbb{C}$. Let D_1 be the Siegel domain of type II associated with the above V and F . It is known in [14] that D_1 is a non-symmetric homogeneous bounded domain in $\mathbb{C}^4$. We have $\mathfrak{s} = \tilde{\mathfrak{s}}_1 = \mathfrak{sl}(2,\mathbb{R})$ and $\mathfrak{k} = \mathfrak{so}(2)$. Using the roots theory of a j-algebra, we can conclude that the universal domain D of D_1 is the Siegel domain of type I over the cone V_5 , where

$$V_5 = \left\{ (a_1,a_2,a_3,x,y) \in \mathbb{R}^5 \ ; \ \begin{pmatrix} a_1 & x & y \\ x & a_2 & 0 \\ y & 0 & a_3 \end{pmatrix} > 0 \right\} .$$

2) Let $(\mathfrak{l},j,\rho)$ be a 2m-dimensional symplectic vector space. Let R be a 1-dimensional subspace of $\mathfrak{l}$, and W be the subspace consisting of all the elements $x \in \mathfrak{l}$ satisfying $\rho(x,R+jR)=0$. Then $\mathfrak{l}$ can be represented as $\mathfrak{l}=jR+R+W$ (a vector space direct sum). Let r_o be a non-zero element in R . We define bracket relations in $\mathfrak{l}$ by

$$[jr_o,r_o] = r_o , \quad [jr_o,u] = \frac{1}{2} u , \quad u \in W ,$$

$$[r_o,u] = 0 ,$$

$$[u,v] = \frac{\rho(u,v)}{\rho(jr_o,r_o)} r_o , \quad u,v \in W .$$

Then $\mathfrak{l}$ is a Lie algebra. Let us define a linear form ω on $\mathfrak{l}$ by $\omega(x)=\rho(jr_o,x)$. Then we have $\rho(u,v)=\omega([u,v])$ for $u,v \in W$. And $\{\mathfrak{l},j,\omega\}$ is a normal j-algebra (cf. [15]), which is j-isomorphic to the Iwasawa j-algebra of the homogeneous bounded domain $D_1 = \{(z_1,\ldots,z_m) \in \mathbb{C}^m : |z_1|^2+\ldots+|z_m|^2<1\}$. Let $p \in \mathfrak{gl}(W)$. Then $[pu,v] + [u,pv] = 0$ if and only if $\rho(pu,v)+\rho(u,pv)=0$. Hence we see that $\mathfrak{s} = \tilde{\mathfrak{s}}_1 = \mathfrak{sp}(W) \cong \mathfrak{sp}(m-1,\mathbb{R})$. It is known that the universal domain D of D_1 is the Siegel upper-half plane $D(H^+(m,\mathbb{R}))$ of degree m .

We will show that the universal domain D of D_1 is determined uniquely by D_1 up to holomorphic equivalence. Let G' be the simply connected Lie group corresponding to $\mathfrak{g}$, and S' and K' be the analytic subgroups of G' corresponding to $\mathfrak{s}$ and $\mathfrak{k}$, respectively. Then by a theorem of Mostow (cf. [5]), S' is closed in G' , and K' is closed in S' , since $\mathfrak{k}$ is a maximal compact subalgebra of $\mathfrak{s}$.

So K' is closed in G'. We consider the simply connected coset space G'/K'. Let Δ be the maximal normal subgroup of G' contained in K'. Since $\{\mathfrak{g},\mathfrak{k}\}$ is effective, Δ is a discrete subgroup of G'. We define the groups G and K as $G=G'/\Delta$ and $K=K'/\Delta$. Then G acts on G'/K' effectively and we have $G'/K'=G/K$.

Lemma 5.3. G/K is a homogeneous Kähler manifold.

Proof. Since the j-algebra $\{\mathfrak{g},\mathfrak{k},j,\omega\}$ satisfies ji)-jiv) and K is connected, G/K is a homogeneous complex manifold. The linear form ω is extended to the left-invariant 1-form on G, which is also denoted by ω. Let $X\in\mathfrak{k}$ and $Y\in\mathfrak{g}$. Then by jv) we have

$$(L_X\omega)(Y) = X\cdot\omega(Y) - \omega([X,Y]) = 0 ,$$

where L_X is the Lie derivative along X. This implies that ω is the right-invariant under K. Hence the 2-form $\rho=d\omega$ is left-invariant under G and right-invariant under K. Furthermore $\rho\equiv 0 \pmod{\mathfrak{k}}$. Hence, there exists a closed invariant 2-form ρ_1 on G/K satisfying $\pi^*\rho_1=\rho$, π being the natural projection of G to G/K. Let J be the complex structure on G/K. For two vector fields X,Y on G/K we put $h(X,Y)=\rho_1(X,JY)$. Then h is a G-invariant Kähler metric on G/K.

QED

Proposition 5.4. Let D be the universal domain of D_1. Then D is holomorphically equivalent to G/K.

Proof. Let T be the subgroup of G generated by $\mathfrak{t}$. Then the orbit $T\cdot o$ is open in G/K, o being the origin of G/K. Hence, by Lemma 5.3 and a result of Kobayashi-Nomizu [10], we have $T\cdot o=G/K$. So we get $G/K=T/T\cap K$, $T\cap K$ being a discrete subgroup of T. Since G/K is simply connected, $T\cap K=(1)$ and T is simply connected. On the other hand, the simply connected group T acts also on D simply transitively. Let $z_o\in D$ be the point corresponding to the point $(ir,0)$ of the Siegel domain of type II holomorphically equivalent to D, where r is given in Theorem 3.11. We denote by ϖ and ϖ' the natural projections defined by $\varpi(t)=t\cdot z_o$ and $\varpi'(t)=t\cdot o$ for $t\in T$. Then it follows that $\varpi'\cdot\varpi^{-1}$ is a holomorphic homeomorphism of D onto G/K.

QED

Corollary 5.5. $G = K \cdot T$ (semi-direct)

Proposition 5.6. *The universal domain* D *of* D_1 *is uniquely determined by* D_1 *up to holomorphic equivalence*.

Proof. Let $\mathfrak{f}_1$ and $\mathfrak{f}_1'$ be two Iwasawa j-algebras of D_1. Then by Proposition 5.2 in [7], $\mathfrak{f}_1'$ is j-isomorphic to $\mathfrak{f}_1$. Let $\mathfrak{f}_1 = jR+R+W$ and $\mathfrak{f}_1' = jR'+R'+W'$ be the decompositions in Theorem 3.11. Then the above j-isomorphism preserves these decompositions (cf. Proposition 5.1 in [7]). Hence W and W' are mutually isomorphic as symplectic vector spaces, from which it follows that the universal j-algebras $\mathfrak{g}$ and $\mathfrak{g}'$ corresponding to $\mathfrak{f}_1$ and $\mathfrak{f}_1'$ are mutually j-isomorphic. Let $\mathfrak{g} = \mathfrak{f}_1 + \mathfrak{s}$ and $\mathfrak{g}' = \mathfrak{f}_1' + \mathfrak{s}'$ be the decompositions given in §4, where $\mathfrak{s}$ and $\mathfrak{s}'$ are the semi-simple j-subalgebras of $\mathfrak{g}$ and $\mathfrak{g}'$, respectively.

Let $\mathfrak{f}_{\mathfrak{s}}$ (resp. $\mathfrak{f}_{\mathfrak{s}}'$) be a maximal $\mathbb{R}$-triangular subalgebra of $\mathfrak{s}$ (resp. $\mathfrak{s}'$). Then by Lemma 5.1, $\mathfrak{f} = \mathfrak{f}_1 + \mathfrak{f}_{\mathfrak{s}}$ (resp. $\mathfrak{f}' = \mathfrak{f}_1' + \mathfrak{f}_{\mathfrak{s}}'$) is a normal j-subalgebra of $\mathfrak{g}$ (resp. $\mathfrak{g}'$). Let $D(\mathfrak{f})$ (resp. $D(\mathfrak{f}')$) be the homogeneous bounded domain corresponding to $\mathfrak{f}$ (resp. $\mathfrak{f}'$). Then by Proposition 5.4, $D(\mathfrak{f})$ is represented as the complex coset space G/K. On the other hand $D(\mathfrak{f}')$ is also represented as G/K, since $\mathfrak{g}'$ is j-isomorphic to $\mathfrak{g}$. This implies that $D(\mathfrak{f}')$ is holomorphically equivalent to $D(\mathfrak{f})$, and the universal domain $D=D(\mathfrak{f})$ is uniquely determined by D_1 up to holomorphic equivalence.

QED

Theorem B. (Kaneyuki [7]) *Let* D *be a homogeneous bounded domain in* $\mathbb{C}^n$, G_h *the identity component of the full automorphism group of* D, *and* T *be the Iwasawa subgroup of* G_h. *Then there exists a faithful representation* τ *of* G_h *such that* $\tau(G_h)$ *is the identity component of a real algebraic group*. *Furthermore* $\tau(T)$ *is a maximal* $\mathbb{R}$*-triangular subgroup of* $\tau(G_h)$.

On the properties of the group G we have

Proposition 5.7. 1) G *is isomorphic to the identity component of a real algebraic group via the representation* τ.

2) K *is a maximal compact subgroup of* G, *and* T *is a maximal* $\mathbb{R}$*-triangular subgroup of* G *via* τ.

3) $G = T_1 \cdot S$, (semi-direct) ,

<u>where</u> S <u>is the analytic subgroup of</u> G <u>generated by</u> $\mathfrak{s}$.

4) G <u>is centerless</u>.

<u>Proof.</u> We use the notations in Theorem B.

1) It is enough to prove that the Lie algebra $\tau(\mathfrak{g})$ is algebraic. $\mathfrak{t}$ is the Iwasawa j-algebra of D . Hence by Theorem B, $\tau(\mathfrak{t})$ is a maximal $\mathbf{R}$-triangular subalgebra in $\tau(\mathfrak{g}_h)$. Furthermore $\tau(\mathfrak{g}_h)$ is algebraic by Theorem B. Hence $\tau(\mathfrak{t})$ is algebraic. On the other hand $\tau(\mathfrak{s})$ is semi-simple, so $\tau(\mathfrak{s})$ is algebraic. Therefore $\tau(\mathfrak{g})$, spanned by $\tau(\mathfrak{t})$ and $\tau(\mathfrak{s})$, is algebraic.

2) Let K_h be the isotropy subgroup of G_h at the origin o in D . Then K_h is compact in G_h . By 1), G is closed in G_h , and $K = K_h \cap G$. Hence K is a compact subgroup of G . On the other hand $D = G/K$ is a cell, which implies that K is a maximal compact subgroup of G . The other statement is an immediate consequence of the fact that $\tau(\mathfrak{t})$ is maximal $\mathbf{R}$-triangular in $\tau(\mathfrak{g})$.

3) Let T_1 and $T_{\mathfrak{s}}$ be the analytic subgroups of T generated by $\mathfrak{t}_1$ and $\mathfrak{t}_{\mathfrak{s}}$, respectively. Since T is simply connected and solvable, T_1 and $T_{\mathfrak{s}}$ are closed and simply connected subgroups of T . Hence

$$T = T_1 \cdot T_{\mathfrak{s}} \qquad \text{(semi-direct)}$$

So we get

$$G = K \cdot T = KT_1 T_{\mathfrak{s}} = T_1 K T_{\mathfrak{s}} = T_1 \cdot S .$$

4) is a special case of Theorem 3.1 in [7] .

QED

Now we consider the fibrings of homogeneous bounded domains. Let $\mathfrak{t}_o$ be the Iwasawa j-algebra of a homogeneous bounded domain D_o , $\mathfrak{t}_1$ a j-ideal of $\mathfrak{t}_o$, and $\mathfrak{t}_o = \mathfrak{t}_1 + \mathfrak{t}_2$ be the decomposition given by (#) in §4. Let $D_i (i=1,2)$ be the homogeneous bounded domain corresponding to $\mathfrak{t}_i (i=1,2)$. Let T_o be the simply connected Lie group corresponding to $\mathfrak{t}_o$ and $T_i (i=1,2)$ be the analytic subgroup of T_o generated by $\mathfrak{t}_i (i=1,2)$. Then T_1 and T_2 are simply connected and closed in T_o .

We have

$$T_o = T_1 \cdot T_2 \qquad \text{(semi-direct)} \qquad (*)$$

Each T_i acts on D_i holomorphically and simply transitively $(i=0,1,2)$. Let $o \varepsilon D_o$ and $o_2 \varepsilon D_2$ be the points corresponding to the origins in the realizations of D_o and D_2 as Siegel domains. Each point $y \varepsilon D_2$ can be written uniquely as $y = t \cdot o_2$, where $t \varepsilon T_2$. Let us define the map ι of D_2 to D_o by putting $\iota(y) = t \cdot o$. Then ι is a holomorphic imbedding of D_2 into D_o , since $\mathfrak{t}_2$ is a j-subalgebra of $\mathfrak{t}_o$. D_2 will be always identified with its image by ι . For each $x \varepsilon D_o$ there exists one and only one $t_o \varepsilon T_o$ such that $x = t_o \cdot o$. By $(*)$, t_o can be written uniquely as $t_o = t_1 \cdot t_2$, where $t_1 \varepsilon T_1$, $t_2 \varepsilon T_2$. We define the map π_o of D_o to D_2 by

$$\pi_o(x) = t_2 \cdot o \ .$$

Then π_o is a holomorphic map of D_o onto D_2 and of maximal rank. In fact we have the following commutative diagram:

$$\begin{array}{ccc} T_o/T_1 \cong T_2 & \xleftarrow{\ \alpha\ } & T_o \\ \downarrow & & \downarrow \\ D_2 = T_2 \cdot o & \xleftarrow{\ \pi_o\ } & T_o \cdot o = D_o \end{array}$$

where α is the natural homomorphism and the vertical arrows are the natural maps. For each $y \varepsilon D_2$ the fibre $\pi_o^{-1}(y)$ coincides with the orbit $T_1 \cdot y$ and so $\pi_o^{-1}(y)$ is holomorphically equivalent to D_1 .

The symmetric bounded domain S/K is denoted by $D(\mathfrak{S})$. $D(\mathfrak{S})$ can be holomorphically imbedded in the universal domain $D = G/K$ as the S-orbit of the origin. Let $x = g \cdot o \,\varepsilon D$, where $g \varepsilon G$. The element g can be written uniquely as $g = t_1 \cdot s$, where $t_1 \varepsilon T_1$, $s \varepsilon S$. We define the map π of D to $D(\mathfrak{S})$ by

$$\pi(x) = sK \ .$$

Then $\pi(x)$ is determined uniquely by x and independent of the choice of $g \varepsilon G$ satisfying $x = g \cdot o$. The map π is a holomorphic map of D onto to $D(\mathfrak{S})$ and of maximal rank. Each fibre $\pi^{-1}(y)$ $(y \varepsilon D(\mathfrak{S}))$ coincides with the orbit $T_1 \cdot y$ which is holomorphically equivalent to D_1 .

Since T_2 is simply connected, the classifying j-homomorphism

λ of $\mathfrak{t}_2$ into $\mathfrak{s}$ (cf. Theorem 4.13) induces the homomorphism of T_2 into S, which is denoted again by λ. Let $y=t\cdot o\in D_2$, $t\in T_2$. Let us define the map μ of D_2 to $D(\mathfrak{s})$ by

$$\mu(y) = \lambda(t)K .$$

Then μ is holomorphic. On the other hand λ is extended to the homomorphism $\tilde{\lambda}$ of $\mathfrak{t}_o$ into $\mathfrak{g}$ by putting $\lambda(t_1)=t_1$ for each $t_1\in\mathfrak{t}_1$. The homomorphism $\tilde{\lambda}$ induces the homomorphism of T_o into G, which is also denoted by $\tilde{\lambda}$. Let $x=t_o\cdot o\in D_o$, $t_o\in T_o$. We define the map $\tilde{\mu}$ of D_o into D by

$$\tilde{\mu}(x) = \tilde{\lambda}(t_o)K .$$

Then $\tilde{\mu}$ is holomorphic and the restriction of $\tilde{\mu}$ to D_2 is μ. We have thus the following

<u>Proposition 5.8.</u> <u>The following diagram is commutative:</u>

$$\begin{array}{ccc} D_o & \xrightarrow{\ \tilde{\mu}\ } & D \\ \downarrow{\scriptstyle \pi_o} & & \downarrow{\scriptstyle \pi} \\ D_2 & \xrightarrow{\ \mu\ } & D(\mathfrak{s}) \end{array}$$

<u>And the action of</u> T_1 <u>commutes with</u> $\tilde{\mu}$.

<u>Definition 5.9.</u> The homogeneous bounded domain $D(\mathfrak{s})$ is called the <u>classifying domain</u> of D_1 ; the map μ is called the <u>classifying map</u>. If $\dim D(\mathfrak{s})>0$, then we say that D_1 has the <u>non-trivial</u> classifying domain.

We remark here that there is a large class of homogeneous bounded domains with non-trivial classifying domains.

§6. A Generalization of Borel Imbeddings

Let M_1 be a connected complex homogeneous space whose universal covering manifold is holomorphically equivalent to a bounded domain D in $\mathbb{C}^n$. Then by Kobayashi [9] M_1 is a hyperbolic manifold; the group Aut M_1 of all holomorphic automorphisms is a Lie group with respect to the compact-open topology and its isotropy subgroups are compact. Let G be the identity component of Aut M_1 and K be the isotropy subgroup of G at a point $o \varepsilon M_1$. G acts on M_1 transitively and M_1 can be represented as G/K ; K is a compact subgroup of G.

Let $\widetilde{G}'$ be the universal covering group of G and $\widetilde{K}'$ be the (closed) analytic subgroup of $\widetilde{G}'$ generated by the Lie algebra $\mathfrak{k}$ of K. The simply connected covering manifold $\widetilde{G}'/\widetilde{K}'$ of M_1 has the $\widetilde{G}'$-invariant complex structure such that the natural projection ϖ of $\widetilde{G}'/\widetilde{K}'$ onto M_1 is holomorphic. By the uniqueness of simply connected covering spaces, D is holomorphically equivalent to $\widetilde{G}'/\widetilde{K}'$. Let π be the covering homomorphism of $\widetilde{G}'$ onto G and Δ_o be the subgroup consisting of all the elements $g \varepsilon \widetilde{G}'$ such that $g \cdot y = y$ for each $y \varepsilon D$. Then Δ_o is the (unique) maximal normal subgroup of $\widetilde{G}'$ contained in $\widetilde{K}'$. Since G acts on M_1 effectively, Δ_o is a discrete normal subgroup of $\widetilde{G}'$. Let us consider the factor groups $\widetilde{G} = \widetilde{G}'/\Delta_o$ and $\widetilde{K} = \widetilde{K}'/\Delta_o$: D is represented as the complex coset space $\widetilde{G}/\widetilde{K}$. $\widetilde{G}$ acts on D effectively. M_1 can be represented as $M_1 = \widetilde{G}'/\pi^{-1}(K)$. Let Δ be the subgroup consisting of all the elements $g \varepsilon \widetilde{G}'$ such that $g \cdot x = x$ for each $x \varepsilon M_1$. Then $\Delta \supset \Delta_o$. Hence there exists the natural (covering) homomorphism π' of $\widetilde{G}$ onto G : M_1 can be represented as $M_1 = \widetilde{G}/\pi'^{-1}(K)$, and $\widetilde{G}$ acts on M_1 holomorphically.

Let G_h be the identity component of the full automorphism group $G_h(D)$ of D. Then $\widetilde{G} \subset G_h$.

$$\begin{array}{ccccc} & & \widetilde{G}' & & \\ & \swarrow & & \searrow \pi & \\ \widetilde{G}'/\Delta_o = \widetilde{G} & \xrightarrow{\ \pi'\ } & & & G = \widetilde{G}'/\Delta \end{array}$$

Lemma 6.1. The group $\widetilde{G}$ is the unique maximal connected Lie subgroup of G_h whose action is projectable on M_1.

Proof. $\widetilde{G}$ is itself a Lie subgroup of G_h. Let $\widetilde{G}_1$ be the identity component of the Lie subgroup of G_h consisting of all the elements of

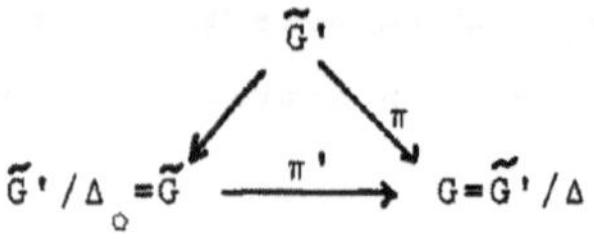

G_h whose actions are projectable on M_1. Then $\tilde{G}_1 \supset \tilde{G}$. Suppose $\tilde{G}_1 \supsetneq \tilde{G}$. First we need to show that $\tilde{G}_1$ acts on M_1 holomorphically. Let I and $\tilde{I}$ be the complex structures on M_1 and D, respectively, and let $g \varepsilon \tilde{G}_1$. We denote by g_{*x} the differential of the map induced by g at x. Let y be an arbitrary point in D and let $\varpi(y)=x$. Since the projection ϖ is holomorphic and $g \cdot \varpi = \varpi \cdot g$, we have

$$g_{*x} I_x \varpi_{*y} = g_{*x} \varpi_{*y} \tilde{I}_y = \varpi_{*y} g_{*y} \tilde{I}_y = \varpi_{*y} \tilde{I}_{g\cdot y} g_{*y}$$

$$= I_{g\cdot x} \varpi_{*y} g_{*y} = I_{g\cdot x} g_{*x} \varpi_{*y} .$$

The differential ϖ_{*y} is a linear isomorphism. So we have $g_{*x} I_x = I_{g\cdot x} g_{*x}$, which implies that the action of g is holomorphic on M_1.

Let N be the closed normal subgroup consisting of all the elements $g \varepsilon \tilde{G}_1$ such that $g \cdot x = x$ for all $x \varepsilon M_1$. Let ϕ be the natural projection of $\tilde{G}_1$ onto $\tilde{G}_1/N$. $N \cap \tilde{G}$ coincides with the kernel of the covering homomorphism π' of $\tilde{G}$ to G. Hence we get $\phi(\tilde{G})=\tilde{G}/N \cap \tilde{G} = \tilde{G}/\mathrm{Ker}\pi' = G$. On the other hand it follows from the above argument that $\phi(\tilde{G}_1)=\tilde{G}_1/N \subset G$. Hence we get $\phi(\tilde{G}_1)=G$. Since $\dim \tilde{G}_1 > \dim \tilde{G}$, N is not discrete. Let $\mathfrak{n}$ be the Lie algebra of N and X be a non-zero element in $\mathfrak{n}$. Let X^* (resp. X^{**}) be the vector field on D (resp. M_1) induced by the one-parameter subgroup $\exp tX$. Then X^{**} is ϖ-related to X^*. The action of N on M_1 is trivial, so $X^{**}=0$, which implies $X^*=0$. $\tilde{G}_1$ acts on D effectively. So it follows that $X=0$, which is a contradiction. We have thus proved the lemma.

QED

Let Γ be the fundamental group of M_1. Then Γ acts on D holomorphically, and so Γ is a subgroup of $G_h(D)$. As is proved in the appendix in [7] we can take the representation $\mathrm{Ad}\bullet\mathrm{Ad}$ as the faithful representation τ of G_h in Theorem B, §4. $\mathrm{Ad}\bullet\mathrm{Ad}$ is also a representation of $G_h(D)$. For brevity $\mathrm{Ad}\bullet\mathrm{Ad}$ will be denoted by ρ. For an arbitrary Lie group G its identity component will be denoted by G^o.

<u>Lemma 6.2.</u> <u>Let</u> $Z_{G_h(D)}(\Gamma)$ <u>be the centralizer of</u> Γ <u>in</u> $G_h(D)$. <u>Then</u> $Z_{G_h(D)}(\Gamma)^o$ <u>is isomorphic to the identity component of a real algebraic group via the representation</u> ρ.

Proof. The group $Z_{G_h(D)}(\Gamma)$ is closed subgroup of $G_h(D)$. So it is a regularly imbedded Lie subgroup of $G_h(D)$. Note that $Z_{G_h(D)}(\Gamma)^\circ = Z_{G_h}(\Gamma)^\circ$. First we will show that $\rho(Z_{G_h}(\Gamma))$ coincides with the centralizer $Z_{\rho(G_h)}(\rho(\Gamma))$ of $\rho(\Gamma)$ in $\rho(G_h)$. Let $\rho(g)\varepsilon Z_{\rho(G_h)}(\rho(\Gamma))$, where $g\varepsilon G_h$. Then $\rho(\gamma g\gamma^{-1})=\rho(g)$ for all $\gamma\varepsilon\Gamma$. The element $\gamma g\gamma^{-1}$ is in G_h and ρ is faithful on G_h. Hence it follows that $\gamma g\gamma^{-1}=g$ for all $\gamma\varepsilon\Gamma$, which implies $\rho(g)\varepsilon\rho(Z_{G_h}(\Gamma))$, or $Z_{\rho(G_h)}(\rho(\Gamma))\subset\rho(Z_{G_h}(\Gamma))$. The converse inclusion is trivial.

Let us denote by $Z(\rho(\Gamma))$ the centralizer of $\rho(\Gamma)$ in the full linear group. Then $Z_{\rho(G_h)}(\rho(\Gamma))=\rho(G_h)\cap Z(\rho(\Gamma))$. Therefore $Z_{\rho(G_h)}(\rho(\Gamma))$ is closed subgroup of $\rho(G_h)$. In particular $\rho(Z_{G_h}(\Gamma))$ has the structure of a regularly imbedded Lie subgroup of $\rho(G_h)$. Hence ρ is an one-to-one continuous homomorphism of $Z_{G_h}(\Gamma)$ onto $\rho(Z_{G_h}(\Gamma))$. Since $Z_{G_h}(\Gamma)$ has a countable base, ρ is an isomorphism of $Z_{G_h}(\Gamma)$ onto $\rho(Z_{G_h}(\Gamma))$ as Lie groups. In particular $\rho(Z_{G_h}(\Gamma)^\circ)=\rho(Z_{G_h}(\Gamma))^\circ$. By Theorem B, §4, $\rho(G_h)$ is the identity component of a real algebraic group, which is denoted by $\hat{G}$. It is easy to see that $(\rho(G_h)\cap Z(\rho(\Gamma)))^\circ$ coincides with $(\hat{G}\cap Z(\rho(\Gamma)))^\circ$, the identity component of the algebraic group $\hat{G}\cap Z(\rho(\Gamma))$. From these arguments we conclude that $Z_{G_h(D)}(\Gamma)^\circ$ is isomorphic to $(\hat{G}\cap Z(\rho(\Gamma)))^\circ$ via ρ.

QED

Proposition 6.3. Let M_1 be a complex homogeneous space. Suppose that the universal covering manifold of M_1 is holomorphically equivalent to a bounded domain D in $\mathbb{C}^n$. Then M_1 itself is holomorphically equivalent to D.

Proof. We will use previous notations. D is represented as $D=\tilde{G}/\tilde{K}$ $=G_h/K_h$, where K_h and $\tilde{K}$ are isotropy subgroups at the same point in M_1. M_1 is considered to be the factor space $\Gamma\backslash D$. Let $\mathfrak{g}_h$ be the

Lie algebra of G_h , which is regarded as the Lie algebra of all complete holomorphic vector fields on D . Let $\mathfrak{g}$ be the Lie algebra of the group $\tilde{G}$. Then by Lemma 6.1 $\mathfrak{g}$ consists of all projectable complete holomorphic vector fields on D , which are characterized as Γ-invariant complete holomorphic vector fields. Hence $\mathfrak{g}$ coincides with the subalgebra of $\mathfrak{g}_h$ consisting of all the elements X satisfying $(\mathrm{Ad}\gamma)X=X$ for all $\gamma\in\Gamma$, which implies that $\tilde{G}=Z_{G_h(D)}(\Gamma)^{\circ}$. By Lemma 6.2 $\rho(\tilde{G})$ is the identity component of a real algebraic group. In particular $\tilde{G}$ is closed in G_h . Since $\tilde{K}=\tilde{G}\cap K_h$ and K_h is a compact subgroup of G_h , $\tilde{K}$ is compact in $\tilde{G}$. The homogeneous bounded domain D is a cell. So $\tilde{K}$ is a maximal compact subgroup of $\tilde{G}$. By a theorem of Vinberg [19] , there exists a connected subgroup T of $\tilde{G}$ such that $\rho(T)$ is a maximal $\mathbb{R}$-triangular subgroup of $\rho(\tilde{G})$ and such that $\rho(\tilde{G})=\rho(\tilde{K})\cdot\rho(T)$ (semi-direct). Theorem B, §4, implies that T is the Iwasawa subgroup of G_h . Hence by Theorem 3.1 in [7] $\tilde{G}$ is centerless and so $\tilde{G}$ is isomorphic to G . And $\tilde{K}$ is isomorphic to K° . Hence K° is a maximal compact subgroup of G . K is a compact subgroup containing K° , which implies that K is connected. And we have $D=\tilde{G}/\tilde{K}=G/K=M_1$.

QED

Let D be a homogeneous bounded domain in $\mathbb{C}^n$, and K_h be the isotropy subgroup of G_h at $o\in D$. Let T be the Iwasawa subgroup of G_h . Then we have

$$G_h = K_h\cdot T \qquad \text{(semi-direct)} .$$

Let $\mathfrak{k}_h$ and $\mathfrak{t}$ be the Lie algebras of K_h and T , respectively. Then the pair $\{\mathfrak{g}_h, \mathfrak{k}_h\}$ has the j-algebra structure together with the collection (j) and the linear form ω ; the triple $\{\mathfrak{t}, j_t, \omega\}$ is the Iwasawa j-algebra of D . We denote the complexifications of $\mathfrak{g}_h, \mathfrak{k}_h, \mathfrak{t}$ by $\mathfrak{g}_h^C, \mathfrak{k}_h^C, \mathfrak{t}^C$, respectively. Let us define the complex subalgebras $\mathfrak{g}_h^+, \mathfrak{g}_h^-, \mathfrak{t}^+$ and $\mathfrak{t}^-$ of $\mathfrak{g}_h^C$ by

$$\mathfrak{g}_h^{\pm} = \{x\mp ijx\in \mathfrak{g}_h^C \ ; \ x\in \mathfrak{g}_h,\ j\in(j)\}$$

$$\mathfrak{t}^{\pm} = \{x\mp ij_t x\in \mathfrak{t}^C : x\in \mathfrak{t}\} .$$

Then we have

$$\mathfrak{g}_h^C = \mathfrak{g}_h^+ + \mathfrak{g}_h^-$$

$$\mathfrak{g}_h^+ \cap \mathfrak{g}_h^- = \mathfrak{k}_h^C$$

$$\mathfrak{t}^C = \mathfrak{t}^+ + \mathfrak{t}^-$$

$$\mathfrak{t}^+ \cap \mathfrak{t}^- = (0)$$

$$\overline{\mathfrak{g}_h^+} = \mathfrak{g}_h^- , \overline{\mathfrak{t}^+} = \mathfrak{t}^- ,$$

where the bars mean the conjugation of $\mathfrak{g}_h^C$ with respect to $\mathfrak{g}_h$. Also we have

$$\mathfrak{g}_h^C = \mathfrak{k}_h^C + \mathfrak{t}^+ + \mathfrak{t}^- \qquad \text{(semi-direct)} \qquad (*)$$

<u>Lemma 6.4.</u> $\mathfrak{t}^+ \cap \mathfrak{g}_h^- = \mathfrak{t}^- \cap \mathfrak{g}_h^+ = (0)$.

<u>Proof.</u> Let $x \in \mathfrak{t}^+ \cap \mathfrak{g}_h^-$. Then x can be written as

$$x = t - ij_t t = y + ijy , \ t \in \mathfrak{t}, \ y \in \mathfrak{g}_h .$$

Hence $t=y$ and $j_t t + jy = 0$. Since $\mathfrak{t}$ is a j-subalgebra of $\mathfrak{g}_h$, j_t is congruent to j mod $\mathfrak{k}_h$. So we have

$$0 = j^2(t+y) \equiv -(t+y) \bmod \mathfrak{k}_h .$$

Hence $t \in \mathfrak{k}_h \cap \mathfrak{t} = (0)$, which implies $x=0$.

QED

<u>Lemma 6.5.</u>

$$\mathfrak{g}_h^+ = \mathfrak{k}_h^C + \mathfrak{t}^+ \qquad \text{(semi-direct)}$$

$$\mathfrak{g}_h^- = \mathfrak{k}_h^C + \mathfrak{t}^- \qquad \text{(semi-direct)}$$

<u>Proof.</u> Since $\mathfrak{t}$ is a j-subalgebra of $\mathfrak{g}_h$, we have $\mathfrak{t}^\pm \subset \mathfrak{g}_h^\pm$. $\mathfrak{k}_h^C \cap \mathfrak{t}^\pm = (0)$ is valid. It is enough to see that $\mathfrak{g}_h^+ \subset \mathfrak{k}_h^C + \mathfrak{t}^+$. Let $z \in \mathfrak{g}_h^+$. By (*), z can be written in the form

$$z = k + t + t' \qquad k \in \mathfrak{k}_h^C , \ t \in \mathfrak{t}^+ , \ t' \in \mathfrak{t}^- .$$

$k+t$ is in $\mathfrak{k}_h^C + \mathfrak{t}^+ \subset \mathfrak{g}_h^+$. Hence by Lemma 6.4

$$t' = z - (k+t) \in \mathfrak{t}^- \cap \mathfrak{g}_h^+ = (0) .$$

So we get $z = k+t \in \mathfrak{k}_h^C + \mathfrak{s}^+$.

QED

<u>Lemma 6.6.</u>

$$\mathfrak{g}_h \cap \mathfrak{g}_h^- = \mathfrak{k}_h$$
$$\mathfrak{g}_h^C = \mathfrak{g}_h + \mathfrak{g}_h^-$$

<u>Proof.</u> To verify the first one, it is enough to show that $\mathfrak{g}_h \cap \mathfrak{g}_h^- \subset \mathfrak{k}_h$. Let $x \in \mathfrak{g}_h \cap \mathfrak{g}_h^-$. By Lemma 6.5 x can be written as $x = k+t$, where $k \in \mathfrak{k}_h^C$, $t \in \mathfrak{s}^-$. Since $x \in \mathfrak{g}_h$, $x=\bar{x}=\bar{k}+\bar{t}$. Hence $x-\bar{k}=\bar{t} \in \mathfrak{s}^+ \cap \mathfrak{g}_h^-$. By Lemma 6.4 we have $t=0$. Hence $x=k \in \mathfrak{k}_h$. To prove the second one, let us consider the dimension of the right-hand side:

$$\begin{aligned}\dim(\mathfrak{g}_h + \mathfrak{g}_h^-) &= \dim \mathfrak{g}_h + \dim \mathfrak{g}_h^- - \dim(\mathfrak{g}_h \cap \mathfrak{g}_h^-)\\ &= \dim \mathfrak{g}_h + \dim \mathfrak{g}_h^- - \dim \mathfrak{k}_h\\ &= \dim \mathfrak{s}^+ + \dim(\mathfrak{s}^- + \mathfrak{k}_h^C) = \dim \mathfrak{g}_h^C ,\end{aligned}$$

which proves the second one.

QED

The following lemma is essentially due to Tanaka [17] .

<u>Lemma 6.7. Let $\mathfrak{l}$ be the normalizer of $\mathfrak{g}_h^-$ in $\mathfrak{g}_h^C$</u> . <u>Then</u> $\mathfrak{l} = \mathfrak{g}_h^-$.

<u>Proof.</u> By Theorem 3.11 the Iwasawa j-algebra $\mathfrak{s}$ can be written in the form:

$$\mathfrak{s} = j_t R + R + W ,$$

and there exists an element $r \in R$ such that $[j_t a, r] = a$ for all $a \in R$. According to Vinberg, Gindikin and Piatetski-Sapiro [20] the subspaces $j_t R$, R and W are the eigenspaces of the operator $\operatorname{ad} j_t r$ corresponding to the eigenvalue 0 , 1 and $\frac{1}{2}$, respectively.

Suppose that $\mathfrak{l} \supsetneqq \mathfrak{g}_h^-$. Then $\mathfrak{l} \cap \mathfrak{s}^+ \neq (0)$. Let x be a non-zero element in $\mathfrak{l} \cap \mathfrak{s}^+$. x can be written in the form

$$x = (j_t a+b+u) - i j_t(j_t a+b+u) ,$$

where $a, b \in R$, $u \in W$. Let us consider the element $j_t r - ir = -i(r+ij_t r)$ in

$\mathfrak{k}^- \subset \mathfrak{g}_h^-$. Then $[j_t r - ir, x] \in \mathfrak{g}_h^-$. Using Theorem 3.11 and jiv), we have

$$[j_t r-ir,x] = j_t [j_t r,a] - j_t a + [j_t r,b] + [j_t r,u] + i[j_t r,a]$$
$$- ij_t [j_t r,b] + ij_t b - i[j_t r, j_t u] + ia + b .$$

In view of the decomposition into eigenspaces by $\operatorname{ad} j_t r$ mentioned above, we get

$$[j_t r-ir,x] = 2ia + 2b + \frac{1}{2}(u-ij_t u)$$

$$= (b-ij_t b) + i(a-ij_t a) + \frac{1}{2}(u-ij_t u) + (b+ij_t b) + i(a+ij_t a) .$$

The left-hand side and the last two terms of the right-hand side belong to $\mathfrak{g}_h^-$: the first three terms of the right-hand side are in $\mathfrak{k}^+$. Hence it follows that

$$(j_t a+b+\frac{1}{2}u) - ij_t(j_t a+b+\frac{1}{2}u) = 0 .$$

So we get $u=0$ and $j_t a+b=0$. Hence $x=0$, which is a contradiction. So we conclude that $\mathfrak{l} = \mathfrak{g}_h^-$.

QED

We will identify G_h with $\tau(G_h)$, where τ is the faithful representation in Theorem B, §4. Then $\mathfrak{g}_h$ and $\mathfrak{g}_h^C$ are linear Lie algebras. Let G_h^C be the analytic subgroup of the full linear group, corresponding to $\mathfrak{g}_h^C$. Then $G_h \subset G_h^C$. Let L be the complex closed subgroup of G_h^C defined by

$$L = \{g \in G_h^C ; (\operatorname{Ad} g)\mathfrak{g}_h^- = \mathfrak{g}_h^-\} .$$

Then the Lie algebra of L is $\mathfrak{l} = \mathfrak{g}_h^-$ by Lemma 6.7. We define the complex manifold M as $M=G_h^C/L$.

<u>Theorem 6.8.</u> <u>Let</u> D <u>be a homogeneous bounded domain in</u> $\mathbb{C}^n$. <u>Then</u> D <u>is holomorphically imbedded, as the</u> G_h<u>- orbit of the origin</u> o' <u>, in the complex homogeneous space</u> $M=G_h^C/L$.

<u>Proof.</u> Let o be the origin of $D=G_h/K_h$. Let $x=gK_h \in D$, where $g \in G_h$. Let us put $\beta(x)=g\cdot o' \in M$. The map β is well-defined. First we will

prove that β is a holomorphic map of D into M. To prove this, it is enough to show that the differential β_{*o} at o is $\mathbb{C}$-linear. Let π be the linear map of $\mathfrak{g}_h$ onto the tangent space $T_o(D)$ at o which carries $X \in \mathfrak{g}_h$ to the tangent vector of the orbit $\exp tX \cdot o$ at o. Let π' be the linear map of $\mathfrak{g}_h^C$ onto the tangent space $T_{o'}(M)$ defined analogously. Let π_1 (resp. π_1') be the natural projection of $\mathfrak{g}_h$ (resp. $\mathfrak{g}_h^C$) onto $\mathfrak{g}_h/\mathfrak{k}_h$ (resp. $\mathfrak{g}_h^C/\mathfrak{g}_h^-$). Then there exist linear isomorphisms $\pi_2 : \mathfrak{g}_h/\mathfrak{k}_h \to T_o(D)$ and $\pi_2' : \mathfrak{g}_h^C/\mathfrak{g}_h^- \to T_{o'}(M)$ satisfying $\pi=\pi_2 \circ \pi_1$ and $\pi'=\pi_2' \circ \pi_1'$. Let α be the linear map of $\mathfrak{g}_h/\mathfrak{k}_h$ to $\mathfrak{g}_h^C/\mathfrak{g}_h^-$ induced by the inclusion ι of $\mathfrak{g}_h$ into $\mathfrak{g}_h^C$. Then we have the following diagram:

$$\begin{array}{ccccc} T_{o'}(M) & \xleftarrow{\pi_2'} & \mathfrak{g}_h^C/\mathfrak{g}_h^- & \xleftarrow{\pi_1'} & \mathfrak{g}_h^C \\ \uparrow \beta_{*o} & & \uparrow \alpha & & \uparrow \iota \\ T_o(D) & \xleftarrow{\pi_2} & \mathfrak{g}_h/\mathfrak{k}_h & \xleftarrow{\pi_1} & \mathfrak{g}_h \end{array}$$

(with $\pi' : \mathfrak{g}_h^C \to T_{o'}(M)$ along the top and $\pi : \mathfrak{g}_h \to T_o(D)$ along the bottom)

And we have $\alpha \circ \pi_1 = \pi_1' \circ \iota$ and by Lemma 6.6 α is a linear isomorphism. $\mathfrak{g}_h/\mathfrak{k}_h$ and $\mathfrak{g}_h^C/\mathfrak{g}_h^-$ have the complex structures induced by the maps π_2 and π_2' from $T_o(D)$ and $T_{o'}(M)$, respectively. To prove that β_{*o} is $\mathbb{C}$-linear, it is enough to show that α is $\mathbb{C}$-linear and that the left square of the diagram is commutative. Let $X \in \mathfrak{g}_h$. Then for an arbitrary differentiable function f on M we have

$$(\beta_{*o}\pi(X))f = \pi(X)(f \circ \beta) = \lim_{t \to o} \frac{1}{t}\left(f \circ \beta(\exp tX \cdot \beta^{-1}(o')) - f \circ \beta(\beta^{-1}(o'))\right)$$

$$= \lim_{t \to o} \frac{1}{t}(f(\exp tX \cdot o') - f(o')) = (\pi'\iota(X))f ,$$

from which it follows that $\pi_2' \circ \alpha = \beta_{*o} \circ \pi_2$. To verify that α is $\mathbb{C}$-linear, we consider the commutative diagram

$$\begin{array}{ccc}
\mathfrak{g}_h^C/\mathfrak{g}_h^- & \xleftarrow{\ \pi_1'\ } & \mathfrak{t}^+ \\
\uparrow{\scriptstyle \alpha} & & \uparrow{\scriptstyle \alpha_1} \\
\mathfrak{g}_h/\mathfrak{k}_h & \xleftarrow{\ \pi_1\ } & \mathfrak{t}
\end{array}$$

where $\alpha_1(t) = \frac{1}{2}(t-ij_t t)$ for all $t \in \mathfrak{t}$. π_1 and π_1' are $\mathbb{C}$-linear isomorphisms. α_1 being $\mathbb{C}$-linear, α is $\mathbb{C}$-linear. We have thus proved that β is holomorphic. On the other hand we have

$$\beta(D) = G_h o' = G_h/G_h \cap L \ ;$$

$\beta(D)$ has the G_h-invariant complex structure as an open submanifold of M. The Lie algebra of $G_h \cap L$ is $\mathfrak{g}_h \cap \mathfrak{l} = \mathfrak{g}_h \cap \mathfrak{g}_h^- = \mathfrak{k}_h$. So we get $(G_h \cap L)^o = K_h$, which implies that β is the covering map of G_h/K_h onto the complex homogeneous space $G_h/G_h \cap L$. Since β_{*o} is a $\mathbb{C}$-linear isomorphism of $T_o(G_h/K_h)$ onto $T_{o'}(G_h/G_h \cap L) = T_{o'}(M)$, β is a holomorphic covering map of D onto $\beta(D)$. By Proposition 6.3 β is one-to-one and so it is a holomorphic imbedding of D into M.

QED

<u>Remark.</u> Suppose that $D=G_h/K_h$ is a symmetric bounded domain. Then G_h^C is a complex semi-simple Lie group, L is a maximal parabolic subgroup, and $M=G_h^C/L$ is the compact symmetric space dual to D. In this case the map β is the usual Borel imbedding of D.

As a corollary to Theorem 6.8 we have the following Theorem, which is a special case of a result of Kaup-Matsushima-Ochiai:

<u>Theorem 6.9. A homogeneous bounded domain</u> D <u>is imbedded holomorphically and equivariantly in the complex projective N-space</u> $P_N(\mathbb{C})$, <u>where</u> $N=\binom{n}{m}-1$, $n=\dim_C \mathfrak{g}_h^C$, $m=\dim_C \mathfrak{g}_h^-$.

<u>Proof.</u> The proof given here is similar to Borel-Remmert [3]. Let $G_{n,m}$ be the complex Grassman manifold of all m-dimensional subspaces in $\mathfrak{g}_h^C$. G_h^C acts on $G_{n,m}$ by its adjoint representation. From the definition of L it follows that the isotropy subgroup of G_h^C at the

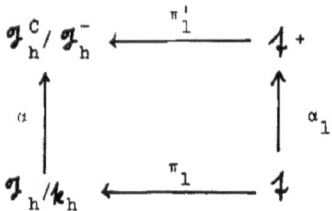

point $\mathfrak{g}_h^- \varepsilon G_{n,m}$ coincides with L. Hence G_h^C/L is imbedded holomorphically in $G_{n,m}$ as the G_h^C-orbit of $\mathfrak{g}_h^- \varepsilon G_{n,m}$. It is well known that $G_{n,m}$ is imbedded in $P_N(\mathbb{C})$ by using the Plücker coordinates. By making the composite of these imbeddings D is imbedded in $P_N(\mathbb{C})$. The action of G_h is induced by the projective transformations.

QED

§7. Cayley Transforms of Universal Domains

Let D_1 be a homogeneous bounded domain in $\mathbb{C}^n$, and $\mathfrak{f}_1 = j_t R + R + W$ be its Iwasawa j-algebra. Let $\{\mathfrak{g}, \mathfrak{k}\}$ be the universal j-algebra of $\mathfrak{f}_1$. Then $\mathfrak{g}$ can be written in the form $\mathfrak{g} = \mathfrak{f}_1 + \mathfrak{s}$ (semi-direct), where $\mathfrak{s}$ is a semi-simple subalgebra of non-compact type containing $\mathfrak{k}$ as a maximal compact subalgebra. The linear endomorphism j on $\mathfrak{g}$ has been defined to be the direct sum of j_t on $\mathfrak{f}_1$ and $j_s = \frac{1}{2}\mathrm{ad}\, j_W$ on $\mathfrak{s}$, where j_W is the restriction of j_t to W (cf. §4). The pair $\{\mathfrak{s}, \mathfrak{k}\}$ together with j_s is a j-subalgebra of $\{\mathfrak{g}, \mathfrak{k}\}$. Let D be the universal domain of D_1. Then, by Proposition 5.4 D is represented as the complex coset space G/K, where G is a connected Lie group corresponding to $\mathfrak{g}$ and K is the analytic subgroup of G generated by $\mathfrak{k}$. G acts on D effectively. Let G_h be the identity component of the full automorphism group of D. Then G is a Lie subgroup of G_h. We can write D as

$$D = G/K = G_h/K_h ,$$

where K_h is the isotropy subgroup of G_h at the origin $o \in G/K$. Let $\{\mathfrak{g}_h, \mathfrak{k}_h, (j), \omega\}$ be the j-algebra of D corresponding to the Lie algebra $\mathfrak{g}_h$ of G_h. This contains $\mathfrak{f}_1$ and $\{\mathfrak{s}, \mathfrak{k}\}$ as j-subalgebras. Let $\mathfrak{f}_1^+$ and $\mathfrak{f}_1^-$ be the subalgebras of $\mathfrak{f}_1^C$ defined in the same way as $\mathfrak{f}^+$ and $\mathfrak{f}^-$ in §6. We have

$$\mathfrak{f}_1^C = \mathfrak{f}_1^+ + \mathfrak{f}_1^-$$

$$\mathfrak{f}_1^+ \cap \mathfrak{f}_1^- = (0)$$

Let

$$\mathfrak{s} = \mathfrak{k} + \mathfrak{p}$$

be the Cartan decomposition associated with the maximal compact subalgebra $\mathfrak{k}$. As is seen in §4, $\mathfrak{p}$ is stable under j_s. The complexification $\mathfrak{p}^C$ is written as $\mathfrak{p}^C = \mathfrak{p}^+ + \mathfrak{p}^-$, where $\mathfrak{p}^+$ and $\mathfrak{p}^-$ are the eigenspaces of j_s corresponding to the eigenvalues i and -i, respectively. Also we have $\mathfrak{p}^{\pm} = \{x \mp i j_s x : x \in \mathfrak{p}\}$, and they are abelian subalgebras of the complexification $\mathfrak{s}^C$ of $\mathfrak{s}$. We get the decomposition

$$\mathfrak{s}^C = \mathfrak{k}^C + \mathfrak{p}^+ + \mathfrak{p}^- ,$$

where $\mathfrak{k}^C$ is the complexification of $\mathfrak{k}$.

In what follows, we will use notations in §6.

Lemma 7.1.

1) $\mathfrak{g}_h = \mathfrak{k}_h + \mathfrak{t}_1 + \mathfrak{p}$ (vector space direct sum)

2) $\mathfrak{g}_h^{\pm} = \mathfrak{k}_h^C + \mathfrak{t}_1^{\pm} + \mathfrak{p}^{\pm}$ (vèctor space direct sum)

3) $\mathfrak{g}_h^C = R^C + W^+ + \mathfrak{p}^+ + \mathfrak{g}_h^-$ (vector space direct sum)

where $W^{\pm} = \{u \mp ij_W u \varepsilon W^C : u \varepsilon W\}$.

4) $[\mathfrak{p}^+, W^+] = [\mathfrak{p}^-, W^-] = 0$

5) $[\mathfrak{p}^+, W^-] \subset W^+$, $[\mathfrak{p}^-, W^+] \subset W^-$

6) $[\mathfrak{p}^+, W] \subset W^+$

7) Let $\mathfrak{u} = R^C + W^+ + \mathfrak{p}^+$. Then $\mathfrak{u}$ is an abelian subalgebra.

8) Let $\mathfrak{u}_1 = R^C + W^+$. Then $[\mathfrak{t}_1, \mathfrak{u}] \subset \mathfrak{u}_1$.

Proof. 1) We know that $\mathfrak{g} = \mathfrak{t}_1 + \mathfrak{k} + \mathfrak{p}$ and that $\mathfrak{g}_h/\mathfrak{k}_h \cong \mathfrak{g}/\mathfrak{k}$ as vector spaces, which implies 1).

2) Since $\mathfrak{t}_1$ and $\mathfrak{p}$ are j-subalgebras of $\mathfrak{g}_h$, we have $\mathfrak{t}_1^{\pm}$, $\mathfrak{p}^{\pm} \subset \mathfrak{g}_h^{\pm}$. 2) is derived by counting the dimensions of the both-hand sides.

3) First we will prove that the right-hand side is a direct sum of the vector spaces. $R^C \cap W^+ = (0)$ and $\mathfrak{p}^+ \cap \mathfrak{g}_h^- = (0)$ are valid. Hence it is enough to see that $(R^C + W^+) \cap (\mathfrak{p}^+ + \mathfrak{g}_h^-) = (0)$. From 2) we see that $\mathfrak{p}^+ + \mathfrak{g}_h^- = \mathfrak{p}^C + \mathfrak{k}_h^C + \mathfrak{t}_1^-$. Let $x \varepsilon (R^C + W^+) \cap (\mathfrak{p}^+ + \mathfrak{g}_h^-)$. x can be written as $x = p + k + t$, where $p \varepsilon \mathfrak{p}^C$, $k \varepsilon \mathfrak{k}_h^C$, $t \varepsilon \mathfrak{t}_1^-$; $x - t \varepsilon \mathfrak{t}_1^C$ and $\mathfrak{t}_1^C \cap (\mathfrak{p}^C + \mathfrak{k}_h^C) = (0)$. Hence $x - t = p + k \varepsilon \mathfrak{t}_1^C \cap (\mathfrak{p}^C + \mathfrak{k}_h^C) = (0)$, which implies that $x = t \varepsilon (R^C + W^+) \cap \mathfrak{t}_1^-$. So x can be written as

$$x = a + ib + u - ij_t u = (a' + j_t b' + u') + ij_t(a' + j_t b' + u'),$$

where $a, b, a', b' \varepsilon R$ and $u, u' \varepsilon W$. Since $R^C, (jR)^C$ and W^C have mutually no intersections, we can conclude $x = 0$. Hence $R^C + W^+ + \mathfrak{p}^+ + \mathfrak{g}_h^-$ is a direct sum. By counting the dimension of $R^C + W^+ + \mathfrak{p}^+ + \mathfrak{g}_h^-$ we get 3).

4) Let $p \in \mathfrak{p}$. Then $pj_W = -j_W p$ (cf. §4). We have

$$p - ij_s p = p - i\cdot\frac{1}{2}[j_W, p] = p - \frac{i}{2}(j_W p - pj_W) = p - ij_W p \ .$$

Let $u \in W$. In view of §4, we have

$$[p - ij_s p,\ u - ij_W u] = [p - ij_W p,\ u - ij_W u]$$

$$= [p,u] - i[p,\ j_W u] - i[j_W p, u] - [j_W p,\ j_W u]$$

$$= pu - ipj_W u - ij_W pu - j_W pj_W u = 0 \ ,$$

which proves 4).

5) We have $[p - ij_s p,\ u + ij_W u] = 2(pu - ij_W pu) \in W^+$.

6) is derived from 5) and 4).

7) One can verify that $[ju, jv] = [u,v]$ for $u, v \in W$, from which it follows that $[W^+, W^+] = 0$. $\mathfrak{p}^+$ and R are abelian, and $[R, \mathfrak{s}] = [R, W] = [\mathfrak{p}^+, W^+] = 0$. So $\mathfrak{n}$ is abelian.

8) $$[\mathfrak{l}_1, \mathfrak{n}] = [jR, R^C] + [jR, W^+] + [jR, \mathfrak{p}^+] + [R, W^+] + [R, \mathfrak{p}^+] + [W, W^+] + [W, \mathfrak{p}^+] \ .$$

Here we know that $[jR, R^C] \subset R^C$, $[W, W^+] \subset R^C$, $[W, \mathfrak{p}^+] \subset W^+$. One can verify that $[jR, W^+] \subset W^+$. Other terms in the right-hand side are zero. So we get 8).

QED

Now we must recall the j-algebra $\{\tilde{\mathfrak{g}}_1, \tilde{\mathfrak{k}}_1\}$ in §4.

<u>Lemma 7.2.</u> <u>The j-algebra</u> $\{\tilde{\mathfrak{g}}_1, \tilde{\mathfrak{k}}_1\}$ <u>is a j-subalgebra of the j-algebra</u> $\{\mathfrak{g}_h, \mathfrak{k}_h, (j), \omega\}$.

<u>Proof.</u> It can be proved that there exists the connected Lie group $\tilde{G}_1$ corresponding to $\tilde{\mathfrak{g}}_1$ satisfying the following two conditions: (i) the coset space $\tilde{G}_1/\tilde{K}_1$ is simply connected, where $\tilde{K}_1$ is the closed analytic subgroup of $\tilde{G}_1$ generated by $\tilde{\mathfrak{k}}_1$; (ii) $\tilde{G}_1$ acts on $\tilde{G}_1/\tilde{K}_1$ effectively. By the same reason as in Lemma 5.3 $\tilde{G}_1/\tilde{K}_1$ is a homogeneous Kähler manifold. Let G^* be the analytic subgroup of $\tilde{G}_1$ generated by

$\mathfrak{g}$. We know in §4 that $\tilde{\mathfrak{g}}_1 = \mathfrak{g} + \tilde{\mathfrak{k}}_1$ and that $\mathfrak{g} \cap \tilde{\mathfrak{k}}_1 = \mathfrak{k}$. Hence, by the same reason as in Proposition 5.4, the G^*-orbit of the origin $o \in \tilde{G}_1/\tilde{K}_1$ coincides with the whole $\tilde{G}_1/\tilde{K}_1$. Hence we have $\tilde{G}_1/\tilde{K}_1 = G^*/G^* \cap \tilde{K}_1$. Since $\tilde{G}_1/\tilde{K}_1$ is simply connected and $\mathfrak{g} \cap \tilde{\mathfrak{k}}_1 = \mathfrak{k}$, the group $G^* \cap \tilde{K}_1$ coincides with the analytic subgroup K^* of $\tilde{G}_1$ generated by $\mathfrak{k}$. On the other hand, by Proposition 5.7 G is centerless. So there exists the covering homomorphism π of G^* to G , and $\pi(K^*) = K$ holds. Hence π induces the holomorphic covering map ϖ of $\tilde{G}_1/\tilde{K}_1 = G^*/K^*$ onto the universal domain $D = G/K$. D being simply connected, ϖ is a diffeomorphism and $\pi^{-1}(K) = K^*$ is valid. So $\operatorname{Ker} \pi$ is contained in K^*. Since G^* acts on D effectively via ϖ , $\operatorname{Ker} \pi$ must be the identity. This implies that $G^* \cong G$ and $K^* \cong K$. Hence D is represented as the complex coset space $\tilde{G}_1/\tilde{K}_1$ on which $\tilde{G}_1$ acts effectively. This implies that $\{\tilde{\mathfrak{g}}_1, \tilde{\mathfrak{k}}_1\}$ is a j-subalgebra of $\{\mathfrak{g}_h, \mathfrak{k}_h\}$.

QED

<u>Lemma 7.3.</u> $j_W \in \mathfrak{g}_h^-$.

<u>Proof.</u> We know in §4 that $j_W \in \tilde{\mathfrak{k}}_1$. And by Lemma 7.2 $\tilde{\mathfrak{k}}_1 \subset \mathfrak{k}_h \subset \mathfrak{g}_h^-$.

QED

We know in Theorem 3.11 that there exists an element r in the abelian ideal R of $\mathfrak{t}_1$ such that $[j_t a, r] = a$ for all $a \in R$. r is uniquely determined in R by the above condition, and is called the <u>unit</u> of R . As is remarked in §6, the decomposition $\mathfrak{t}_1 = j_t R + R + W$ coincides with the decomposition into eigenspaces of the operator $\operatorname{ad} j_t r$; more precisely the subspaces jR , R and W are its eigenspaces corresponding to the eigenvalues 0, 1, and $\frac{1}{2}$, respectively.

Let L be the closed complex subgroup of G_h^C defined in §6. Let N, N_1 and P^+ be the analytic subgroups of G_h^C generated by $\mathfrak{n}$, $\mathfrak{n}_1$ and $\mathfrak{p}^+$, respectively. Then we have

<u>Lemma 7.4.</u> $N \cap L = (1)$.

<u>Proof.</u> Let $g \in N \cap L$. N being abelian, there exists $X = X_1 + X_2 + X_3 \in \mathfrak{n}$ such that $g = \exp X$, where $X_1 \in R^C$, $X_2 \in W^+$ and $X_3 \in \mathfrak{p}^+$. Since $j_t r - ir \in \mathfrak{t}_1^- \subset \mathfrak{g}_h^-$, it follows from the definition of L that $(\operatorname{Ad} g)(j_t r - ir) \in \mathfrak{g}_h^-$.

We have

$$(\mathrm{Ad}\ g)(j_t r - ir) = (\mathrm{Ad}\ \exp X)(j_t r - ir)$$

$$= (j_t r - ir) + [X, j_t r - ir] + \frac{1}{2}[X, [X, j_t r - ir]] + \cdots \quad . (\#)$$

In view of what was remarked before the lemma, we get

$$[X, j_t r - ir] = [X_1 + X_2 + X_3, j_t r - ir]$$

$$= [X_1, j_t r] + [X_2, j_t r] + [X_3, j_t r] = -X_1 - \frac{1}{2}X_2 .$$

Since $-X_1 - \frac{1}{2}X_2 \varepsilon \mathfrak{n}$ and $\mathfrak{n}$ is abelian, the third term $[X, [X, j_t r - ir]]$ of (#) is equal to zero, and consequently

$$(\mathrm{Ad}\ g)(j_t r - ir) = (j_t r - ir) - X_1 - \frac{1}{2}X_2 .$$

The left-hand side and the first term of the right-hand side belong to $\mathfrak{g}_h^-$. So we get $X_1 = X_2 = 0$, and $X = X_3 \varepsilon \mathfrak{g}^+$. Hence $(\mathrm{Ad}\ g)j_W$ is computed as follows:

$$(\mathrm{Ad}\ g)j_W = (\mathrm{Ad}\ \exp X_3)j_W$$

$$= j_W + [X_3, j_W] + \frac{1}{2}[X_3, [X_3, j_W]] + \cdots$$

$$= j_W - 2j_s X_3 + \frac{1}{2}[X_3, -2j_s X_3] + \cdots$$

$$= j_W - 2iX_3$$

By Lemma 7.3 the left-hand side and j_W belong to $\mathfrak{g}_h^-$, which implies that $X = X_3 = 0$, or equivalently g is the identity. So we conclude that $N \cap L = (1)$.

QED

<u>Corollary 7.5.</u> N <u>is simply connected.</u>

<u>Proof.</u> It is enough to see that the exponential map of $\mathfrak{n}$ to N is one-to-one. Let $X \varepsilon \mathfrak{n}$ and suppose that $\exp X$ is the identity. Then $\exp X \varepsilon N \cap L$. The same arguments as in the proof of Lemma 7.4 show that $X = 0$, which implies that exp is one-to-one.

QED

As an immediate consequence of Corollary 7.5 we have

Corollary 7.6. $N = N_1 \cdot P^+$ (direct product) .

Now we will prove that N is a closed subgroup of G_h^C ; we will identify G_h with $\tau(G_h)$ in what follows (cf. §6).

Lemma 7.7. $\mathfrak{t}_1$ is an algebraic subalgebra of $\mathfrak{g}_h$.

Proof. From Lemma 5.1 and the construction of the universal domain D it follows that there exists an Iwasawa j-subalgebra $\mathfrak{t}$ of $\mathfrak{g}_h$ containing $\mathfrak{t}_1$ as a j-ideal such that $\mathfrak{t}$ can be written in the form $\mathfrak{t} = \mathfrak{t}_1 + \mathfrak{t}_s$, where $\mathfrak{t}_s$ is an Iwasawa subalgebra of $\mathfrak{s}$. By Theorem B, §4 $\mathfrak{t}$ is an algebraic subalgebra of $\mathfrak{g}_h$, and so the adjoint representation of $\mathfrak{t}$ is rational. Hence the representation ϕ of $\mathfrak{t}$ on $\mathfrak{t}/\mathfrak{t}_1$ induced by the adjoint representation is also rational (cf. Chevalley [4]). Let $t = t_1 + t_2 \varepsilon \operatorname{Ker} \phi$, where $t_1 \varepsilon \mathfrak{t}_1$ and $t_2 \varepsilon \mathfrak{t}_s$. Then $[t, \mathfrak{t}] \subset \mathfrak{t}_1$, or $[t_1, \mathfrak{t}] + [t_2, \mathfrak{t}] \subset \mathfrak{t}_1$. Since $t_1 \varepsilon \mathfrak{t}_1$ and $\mathfrak{t}_1$ is an ideal of $\mathfrak{t}$, we have $[t_2, \mathfrak{t}] \subset \mathfrak{t}_1$. So we get $[t_2, \mathfrak{t}_s] \subset \mathfrak{t}_1 \cap \mathfrak{t}_s = (0)$, which implies that t_2 is in the center of $\mathfrak{t}_s$. Since the center of $\mathfrak{t}_s$ is zero , we have $t_2 = 0$. Hence $t = t_1 \varepsilon \mathfrak{t}_1$, or $\operatorname{Ker} \phi \subset \mathfrak{t}_1$. The converse inclusion is trivial. Thus $\mathfrak{t}_1$ coincides with the kernel of the rational representation ϕ of the algebraic subalgebra $\mathfrak{t}$. Hence, by Chevalley [4] $\mathfrak{t}_1$ is algebraic.

QED

Lemma 7.8. Let $\mathfrak{c}(\mathfrak{n}_1)$ be the centralizer of $\mathfrak{n}_1$ in the complexification $\mathfrak{t}_1^C$. Then $\mathfrak{c}(\mathfrak{n}_1) = \mathfrak{n}_1$, and $\mathfrak{n}_1$ is an algebraic subalgebra of $\mathfrak{g}_h^C$.

Proof. Let ω_1 be the linear form of the j-algebra $\mathfrak{t}_1$. By the same way as in the proof of Lemma 7.1, 3) one can see that

$$\mathfrak{t}_1^C = \mathfrak{n}_1 + \mathfrak{t}_1^- . \quad \text{(semi-direct)}$$

Hence $\mathfrak{c}(\mathfrak{n}_1)$ can be written as

$$\mathfrak{c}(\mathfrak{n}_1) = \mathfrak{n}_1 + \mathfrak{c}(\mathfrak{n}_1) \cap \mathfrak{t}_1^- .$$

Let $t \varepsilon \mathfrak{c}(\mathfrak{n}_1) \cap \mathfrak{t}_1^-$. Then we can write t as

$$t = (j_t a+b+u) + ij_t(j_t a+b+u) \qquad a,b\varepsilon R \ , \ u\varepsilon W \ .$$

Since the unit $r\varepsilon R$ is in $\mathfrak{n}_1$, we have

$$0 = [t,r] = [j_t a,r] + i[j_t b,r] = a + ib \ ,$$

which implies that $t=u+ij_t u\varepsilon W^-$. Take the element $u-ij_t u\varepsilon W^+\subset\mathfrak{n}_1$. Then we get

$$0 = [t,u-ij_t u] = 2i\,[j_t u,u] \ .$$

Hence $\omega_1([j_t u,u])=0$, which implies that $u=0$. Thus we conclude that $\mathfrak{c}(\mathfrak{n}_1)=\mathfrak{n}_1$. From Lemma 7.7 it follows that $\mathfrak{f}_1^C$ is an algebraic subalgebra of $\mathfrak{g}_h^C$. The centralizer $\tilde{\mathfrak{c}}(\mathfrak{n}_1)$ of $\mathfrak{n}_1$ in the full linear Lie algebra is algebraic, and $\mathfrak{n}_1=\mathfrak{c}(\mathfrak{n}_1)=\mathfrak{f}_1^C\cap\tilde{\mathfrak{c}}(\mathfrak{n}_1)$. Hence $\mathfrak{n}_1$ is algebraic.

QED

<u>Lemma 7.9.</u> <u>Let</u> $\mathfrak{c}(\mathfrak{p}^+)$ <u>be the centralizer of</u> $\mathfrak{p}^+$ <u>in</u> $\mathfrak{f}^C$. <u>Then</u> $\mathfrak{c}(\mathfrak{p}^+)=\mathfrak{p}^+$, <u>and</u> $\mathfrak{p}^+$ <u>is an algebraic subalgebra of</u> $\mathfrak{g}_h^C$.

<u>Proof.</u> Since $\mathfrak{f}^C=\mathfrak{k}^C+\mathfrak{p}^++\mathfrak{p}^-$, we can write $\mathfrak{c}(\mathfrak{p}^+)$ as

$$\mathfrak{c}(\mathfrak{p}^+) = \mathfrak{p}^+ + \mathfrak{c}(\mathfrak{p}^+)\cap(\mathfrak{k}^C+\mathfrak{p}^-) \ .$$

Let $Z=X+Y\varepsilon\,\mathfrak{c}(\mathfrak{p}^+)\cap(\mathfrak{k}^C+\mathfrak{p}^-)$, where $X\varepsilon\,\mathfrak{k}^C$ and $Y\varepsilon\,\mathfrak{p}^-$. Then

$$0 = [Z,\mathfrak{p}^+] = [X,\mathfrak{p}^+] + [Y,\mathfrak{p}^+] \ .$$

Since $[\mathfrak{k}^C,\mathfrak{p}^+]\subset\mathfrak{p}^+$ and $[\mathfrak{p}^-,\mathfrak{p}^+]\subset\mathfrak{k}^C$, we get

$$[X,\mathfrak{p}^+] = [Y,\mathfrak{p}^+] = 0 \ .$$

It is known in [13] that the representation $\mathfrak{k}^C\to\mathrm{ad}_{\mathfrak{p}^+}\mathfrak{k}^C$ is faithful, and so $X=0$. Since $\bar{Y}\varepsilon\,\mathfrak{p}^+$, $[Y,\bar{Y}]=0$ is valid. If we write Y in the form $Y=p+ij_s p$, where $p\varepsilon\,\mathfrak{p}$, then we have

$$2i\,[j_s p,p] = [Y,\bar{Y}] = 0 \ .$$

Let ω_0 be the linear form of the j-algebra $\mathfrak{f}$. The above equality implies that $\omega_0([j_s p,p])=0$. Hence $p\varepsilon\,\mathfrak{k}\cap\mathfrak{p}=(0)$. Thus it follows that $\mathfrak{C}(\mathfrak{g}^+)=\mathfrak{g}^+$. The semi-simple Lie algebra $\mathfrak{f}^C$ is algebraic and the centralizer $\tilde{\mathfrak{C}}(\mathfrak{g}^+)$ of $\mathfrak{g}^+$ in the full linear Lie algebra is algebraic. Since $\mathfrak{C}(\mathfrak{g}^+)=\mathfrak{f}^C\cap\tilde{\mathfrak{C}}(\mathfrak{g}^+)$, $\mathfrak{g}^+$ is algebraic.

Lemma 7.10. $\mathfrak{n}$ is an algebraic subalgebra of $\mathfrak{g}_h^C$ and N is a closed subgroup of G_h^C.

Proof. $\mathfrak{n}$ is spanned by two algebraic subalgebras $\mathfrak{n}_1$ and $\mathfrak{g}^+$. Hence by Chevalley [4] $\mathfrak{n}$ is algebraic. So N is closed in the full linear group. The group G_h^C is also closed in the full linear group, since $\mathfrak{g}_h^C$ is algebraic. Hence N is closed in G_h^C.

QED

Let us put $u^{\pm}=\frac{1}{2}(u\mp ij_t u)$ for $u\varepsilon W$. Since the Lie algebra $R+W$ of the group RW (cf. §2)is a nilpotent algebra of order 2, so is the complexification $(R+W)^C=R^C+W^++W^-$. Hence we have (cf. Lemma 9.6)

$$\exp u^+ \exp u^- = \exp((u^++u^-)+\frac{1}{2}[u^+,u^-]),$$

which implies that

$$\exp u = \exp\frac{1}{2}[u^-,u^+]\exp u^+ \exp u^- . \qquad (*)$$

This formula was originally remarked in Tanaka [17].

Let r be the unit of the abelian ideal R of $\mathfrak{t}_1$ (cf. p.55). In view of Proposition 5.1 in [7], r is uniquely determined by the Iwasawa j-algebra $\mathfrak{t}_1$.

Definition 7.11. Let us define the element $c_1\varepsilon G_h^C$ by $c_1=\exp ir$, where r is the unit of the abelian ideal R of $\mathfrak{t}_1$. c_1 is called the Cayley transformation with respect to the homogeneous bounded domain D_1.

Let T_1' be the analytic subgroup of G_h generated by the subalgebra j_tR. As is seen in §2, the Iwasawa subgroup T_1 corresponding to $\mathfrak{t}_1$ can be written in the form

$$T_1 = T_1'\cdot RW = RW\cdot T_1' \qquad \text{(semi-direct)}$$

Lemma 7.12. Let $h\varepsilon T_1'$. Then $hL=hc_1h^{-1}c_1^{-1}L$.

Proof. First we will show that $c_1 h c_1^{-1} \varepsilon L$; for this we can assume that h is of the form $\exp j_t a$, $a\varepsilon R$. We have

$$c_1 h c_1^{-1} = c_1(\exp j_t a)c_1^{-1} = \exp((\mathrm{Ad}\ c_1)j_t a)$$

$$= \exp((\mathrm{Ad}\ \exp ir)j_t a) = \exp((\exp \mathrm{ad}\ ir)j_t a)$$

$$= \exp(j_t a + [ir, j_t a] + \frac{1}{2}[ir, [ir, j_t a]] + \dots)$$

$$= \exp(j_t a - ia)\varepsilon L ,$$

Since $j_t a - ia \varepsilon \mathfrak{g}_h^-$, which is the Lie algebra of L . Hence $c_1 h^{-1} c_1^{-1} \varepsilon L^{-1} = L$, or $h^{-1} c_1^{-1} L = c_1^{-1} L$. Therefore we have

$$hL = hc_1(c_1^{-1}L) = h c_1(h^{-1}c_1^{-1}L) .$$

QED

Lemma 7.13. $c_1 T_1 L \subset N_1 L$.

Proof. The proof below is similar to that of a lemma in [17] . Since RW is a simply connected nilpotent subgroup and $[R,W]=0$, every element $t_1 \varepsilon T_1$ can be written in the form

$$t_1 = \exp a \exp u \cdot h ,$$

where $a\varepsilon R$, $u\varepsilon W$ and $h\varepsilon T_1'$. Using Lemma 7.12 and the relation (*) before Definition 7.11, we have

$$c_1 t_1 L = c_1 \exp a \exp u \cdot h L = c_1 \exp a \exp u \cdot h\, c_1 h^{-1} c_1^{-1} L$$

$$= c_1 \exp a \exp u \exp i(\mathrm{Ad} h)r \cdot c_1^{-1} L$$

$$= \exp(a + i(\mathrm{Ad} h)r) \exp u \cdot L$$

$$= \exp(i(\mathrm{Ad} h)r + a) \exp \frac{1}{2}[u^-, u^+] \exp u^+ \exp u^- \cdot L$$

$$= \exp(i(\mathrm{Ad} h)r + a + \frac{1}{2}[u^-, u^+] + u^+) \cdot L .$$

Furthermore, since $i(\mathrm{Ad} h)r + a + \frac{1}{2}[u^-, u^+] + u^+ \varepsilon R^c + W^+$, we get $c_1 t_1 L \varepsilon N_1 L$.

QED

With the notations above we have

Theorem 7.14. Let D be the universal domain of a homogeneous bounded domain D_1. Let G_h be the identity component of the full automorphism group of D. Let β be the Borel imbedding of D into the homogeneous space $M=G_h^C/L$, constructed in §6. Let π be the natural projection of G_h^C onto M and ξ be the holomorphic map of the Lie algebra $\mathfrak{n}$ into M defined by $\xi=\pi\circ\exp$. Then we have

1) $c_1\beta(D)\subset\xi(\mathfrak{n})$,
   where c_1 is the Cayley transformation with respect to D_1.

2) ξ is a holomorphic imbedding of $\mathfrak{n}$ and the domain $\xi^{-1}c_1\beta(D)$ in the vector space $\mathfrak{n}$ is holomorphically equivalent to D.

We have the diagram

$$
\begin{array}{ccccc}
D \xrightarrow{\ \beta\ } \beta(D) & & \subset & & M \\
\downarrow c_1 & & & & \downarrow c_1 \\
c_1\beta(D) & \subset & N\cdot o & \subset & M \\
\downarrow \xi^{-1} & & \downarrow \xi^{-1} & & \\
\xi^{-1}c_1\beta(D) & \subset & \mathfrak{n}=R^c+W^++\mathfrak{p}^+ & &
\end{array}
$$

where o is the origin of M.

Proof. 1) We know in Proposition 5.7 that $G=T_1\cdot S$ (semi-direct). On the other hand S is a semi-simple Lie group of non-compact type and K is a maximal compact subgroup of G (cf. §5). So we can write S as

$$S = K\cdot P = P\cdot K ,$$

where $P=\exp\mathfrak{p}$. Since the group G acts on D transitively, we have

$$G_h = G\cdot K_h = T_1\cdot SK_h = T_1PKK_h = T_1PK_h .$$

Let $P^{\pm}$ and K^C be the analytic subgroups of S corresponding to $\mathfrak{p}^{\pm}$ and $\mathfrak{k}^C$, respectively. It is seen from the theory of Harish-Chandra imbeddings that $S\subset P^+K^CP^-$. Hence, considering Lemma 7.1,8), Corollary

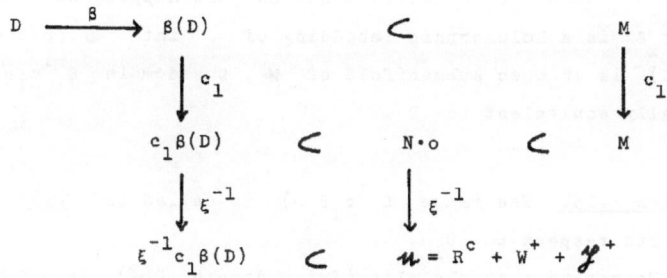

7.6 and Lemma 7.13 we have

$$c_1G_hL = c_1T_1PK_hL = c_1T_1PL \subset c_1T_1P^+K^CP^-L$$

$$= c_1T_1P^+L \subset c_1T_1N_1P^+L = c_1T_1NL$$

$$= c_1NT_1L = Nc_1T_1L \subset NN_1L = NL ,$$

which implies that $c_1\beta(D)=c_1(G_h\cdot o)\subset N\cdot o=\xi(\mathfrak{n})$.

2) By Corollary 7.5 N is simply connected and by Proposition 7.4 π is one-to-one on N . Hence $\xi = \pi\circ\exp$ is a holomorphic homeomorphism of $\mathfrak{n}$ onto the orbit $N\cdot o=N/N\cap L$.

Since the orbit $N\cdot o$ has the same dimension as M , $N\cdot o$ is an open submanifold, and consequently a regularly imbedded submanifold of M . This is derived also from the fact that N is closed in G_h^C (cf. Lemma 7.10). Hence c_1 induces a holomorphic mapping of $\beta(D)$ into $N\cdot o$. So $\xi^{-1}c_1\beta$ is a holomorphic imbedding of D into $\mathfrak{n}$. Since the image $\xi^{-1}c_1\beta(D)$ is an open submanifold of $\mathfrak{n}$, the domain $\xi^{-1}c_1\beta(D)$ is holomorphically equivalent to D .

QED

<u>Definition 7.15.</u> The image $\xi^{-1}c_1\beta(D)$ is called the <u>Cayley transform</u> of D with respect to D_1 .

We remark that the classifying domain $D(\mathfrak{r})$ is naturally imbedded in the universal domain, as is done in §5.

<u>Proposition 7.16.</u> <u>Let</u> $D(\mathfrak{r})$ <u>be the classifying domain of a homogeneous bounded domain</u> D_1 . <u>Then the image</u> $\xi^{-1}c_1\beta(D(\mathfrak{r}))$ <u>coincides with the Harish-Chandra imbedding of the symmetric bounded domain</u> $D(\mathfrak{r})$ <u>into the affine subspace</u> $\mathfrak{p}^+ + ir$ <u>of</u> $\mathfrak{n}$.

<u>Proof.</u> Let us write $D(\mathfrak{r})$ as $D(\mathfrak{r})=S/K$. The group S is contained in $P^+K^CP^-$ and each $g\in S$ can be written uniquely in the form $g=p_1kp_2$, where $p_1\in P^+$, $k\in K^C$, $p_2\in P^-$. There exists one and only one $X\in\mathfrak{p}^+$ such that $p_1=\exp X$. The Harish-Chandra imbedding of $D(\mathfrak{r})$ is the map which carries the coset gK to $X\in\mathfrak{p}^+$. On the other hand we have

$$c_1\beta(gK) = c_1(gL) = c_1(p_1kp_2L) = c_1p_1L$$

$$= \exp(X+ir)\cdot L\epsilon NL \ .$$

Hence we have $\xi^{-1}c_1\beta(gK)=X+ir\epsilon\,\mathfrak{f}^{+}+ir$, which proves the proposition.

QED

§8. Remarks on Harish-Chandra imbeddings

We will use notations in §4 and §7. Let $\{\mathfrak{s}_1, j, \omega_1\}$ be an Iwasawa j-algebra, and let $\mathfrak{s}_1 = jR+R+W$ be the decomposition given in Theorem 3.11. We denote by j_W the restriction of j to W, and define an alternating bilinear form ρ on W as $\rho(u,v)=\omega_1([u,v])$. Then the triple (W, j_W, ρ) is a symplectic vector space. We have denoted by $\mathfrak{sp}(W)$ the Lie algebra of all symplectic endomorphisms on W with respect to ρ, and have defined $\mathfrak{u}(W)$ and $\mathfrak{p}(W)$ to be

$$\mathfrak{u}(W) = \{p\in \mathfrak{sp}(W) : pj_W = j_W p\}$$

$$\mathfrak{p}(W) = \{p\in \mathfrak{sp}(W) : pj_W + j_W p = 0\} .$$

$\mathfrak{u}(W)$ is a subalgebra and we have the decomposition

$$\mathfrak{sp}(W) = \mathfrak{u}(W) + \mathfrak{p}(W) . \qquad \qquad \ldots\ldots\ldots (*)$$

Let us define the hermitian inner product h on W as

$$h(u,v) = \rho(j_W u,v) + i\rho(u,v) .$$

Then we can easily see the following

Lemma 8.1. *Let* $p\in \mathrm{End}\, W$. *Then* $p\in\mathfrak{u}(W)$ *if and only if* $h(pu,v)+h(u,pv)=0$ *for* $u,v\in W$.

Lemma 8.2. *The decomposition* $(*)$ *is a Cartan decomposition*.

Proof. Since ρ is an alternating bilinear form of maximal rank, we can find a base of W such that ρ can be represented as

$$\rho(u,v) = {}^t u \begin{pmatrix} 0 & E \\ -E & 0 \end{pmatrix} v ,$$

where E is the unit matrices of degree $n(\dim W = 2n)$. With respect to this base $\mathfrak{sp}(W)$ is represented as the usual symplectic Lie algebra $\mathfrak{sp}(n,\mathbb{R})$. In particular $\mathfrak{sp}(W)$ is simple. We define the linear endomorphism σ of $\mathfrak{sp}(W)$ as $\sigma|\mathfrak{u}(W)=1$, $\sigma|\mathfrak{p}(W)=-1$. Then , since $[\mathfrak{u}(W), \mathfrak{p}(W)] \subset \mathfrak{p}(W)$ and $[\mathfrak{p}(W), \mathfrak{p}(W)] \subset \mathfrak{u}(W)$, σ is an involutive automorphism. By Lemma 8.1 $\mathfrak{u}(W)$ is compactly imbedded . So

$\{\mathfrak{sp}(W), \mathfrak{u}(W), \sigma\}$ is an effective symmetric triple. With the notations in Helgason [5] we have the decomposition

$$\mathfrak{p}(W) = \mathfrak{p}_o + \mathfrak{p}_+ + \mathfrak{p}_- .$$

The Killing form B of $\mathfrak{sp}(W)$ is positive (resp. negative) definite on $\mathfrak{p}_+$ (resp. $\mathfrak{p}_-$) and $\mathfrak{p}_o$ is an abelian ideal. Since $\mathfrak{sp}(W)$ is simple, $\mathfrak{p}_o$ is reduced to zero. $\mathfrak{p}_-$ generates a compact ideal of $\mathfrak{sp}(W)$, which should be equal to zero. Hence $\mathfrak{p}_- = (0)$ and B is positive definite on $\mathfrak{p}(W)$. The pair $\{\mathfrak{sp}(W), \mathfrak{u}(W)\}$ being effective, B is negative definite on $\mathfrak{u}(W)$. Thus the decomposition $\mathfrak{sp}(W) = \mathfrak{u}(W) + \mathfrak{p}(W)$ is a Cartan decomposition.

QED

By the lemma above $\mathfrak{u}(W)$ is isomorphic to the Lie algebra $\mathfrak{u}(n)$ of the unitary group of degree n. Let j_ρ be the linear endomorphism on W represented as the matrix $\begin{pmatrix} 0 & E \\ -E & 0 \end{pmatrix}$ with respect to the base of W chosen above. Then $j_\rho^2 = -1$ and (W, j_ρ, ρ) is a symplectic vector space. Let us put

$$\mathfrak{u}(\rho) = \{p \varepsilon \mathfrak{sp}(W) : pj_\rho = j_\rho p\}$$

$$\mathfrak{p}(\rho) = \{p \varepsilon \mathfrak{sp}(W) : pj_\rho + j_\rho p = 0\} .$$

Then $\mathfrak{u}(\rho)$ is a maximal compact subalgebra of $\mathfrak{sp}(W)$ and the decomposition

$$\mathfrak{sp}(W) = \mathfrak{u}(\rho) + \mathfrak{p}(\rho)$$

is also a Cartan decomposition. Let Sp(W) be the linear group generated by $\mathfrak{sp}(W)$. By the conjugacy of Cartan decompositions we can find $g \varepsilon Sp(W)$ such that

$$g\, \mathfrak{u}(\rho) g^{-1} = (\mathrm{Ad}\, g)\, \mathfrak{u}(\rho) = \mathfrak{u}(W)$$

$$g\, \mathfrak{p}(\rho) g^{-1} = (\mathrm{Ad}\, g)\, \mathfrak{p}(\rho) = \mathfrak{p}(W) .$$

j_ρ and j_W belong to the centers of $\mathfrak{u}(\rho)$ and of $\mathfrak{u}(W)$, respectively, which are one-dimensional. Ad g carries the center of $\mathfrak{u}(\rho)$ to that of $\mathfrak{u}(W)$. Hence we have

$$g j_\rho g^{-1} = \lambda j_W ,$$

where λ is a real constant. Since $j_\rho^2=j_W^2=-1$, we have $\lambda=\pm 1$, or

$$g j_\rho g^{-1} = \pm j_W .$$

The complexification $\mathfrak{sp}(W)^C$ is decomposed as

$$\begin{aligned}\mathfrak{sp}(W)^C &= \mathfrak{u}(W)^C + \mathfrak{p}(W)^+ + \mathfrak{p}(W)^- \\ &= \mathfrak{u}(\rho)^C + \mathfrak{p}(\rho)^+ + \mathfrak{p}(\rho)^-\end{aligned}$$

by the operators $\frac{1}{2}\mathrm{ad}\, j_W$ and $\frac{1}{2}\mathrm{ad}\, j_\rho$ respectively, where $\mathfrak{p}(W)^\pm$ and $\mathfrak{p}(\rho)^\pm$ are $\pm i$-eigenspaces of $\frac{1}{2}\mathrm{ad}\, j_W$ and $\frac{1}{2}\mathrm{ad}\, j_\rho$, respectively. We have

$$g\, \mathfrak{p}(\rho)^\pm g^{-1} = \mathfrak{p}(W)^\pm ,$$

or

$$g\, \mathfrak{p}(\rho)^\pm g^{-1} = \mathfrak{p}(W)^\mp ,$$

according as $g j_\rho g^{-1}=j_W$ or $g j_\rho g^{-1}=-j_W$, respectively. Let A_g be the inner automorphism by $g \varepsilon Sp(W)$. Then, since $A_g(U(\rho))=U(W)$, A_g induces the diffeomorphism of $Sp(W)/U(\rho)$ onto $Sp(W)/U(W)$, denoted again by A_g. Since the triples $\{\mathfrak{sp}(W), \mathfrak{u}(W), \frac{1}{2}\mathrm{ad}\, j_W\}$ and $\{\mathfrak{sp}(W), \mathfrak{u}(\rho), \frac{1}{2}\mathrm{ad}\, j_\rho\}$ are j-algebras, $D(W)=Sp(W)/U(W)$ and $D(\rho)=Sp(W)/U(\rho)$ have the complex structures induced by $\frac{1}{2}\mathrm{ad}\, j_W$ and $\frac{1}{2}\mathrm{ad}\, j_\rho$, respectively. Then $D(W)$ and $D(\rho)$ are hermitian symmetric spaces of non-compact type.

<u>Lemma 8.3.</u> <u>Suppose</u> $g j_\rho g^{-1}=j_W$. <u>Let</u> κ_1 <u>and</u> κ_2 <u>be the Harish-Chandra imbeddings of</u> $D(W)$ <u>into</u> $\mathfrak{p}(W)^+$ <u>and that of</u> $D(\rho)$ <u>into</u> $\mathfrak{p}(\rho)^+$, <u>respectively</u>. <u>Then the following diagram is commutative</u>,

$$\begin{array}{ccccc} D(W) & \xrightarrow{\kappa_1} & \mathfrak{p}(W)^+ & \xrightarrow{\alpha_1} & \mathfrak{p}(W) \\ \uparrow A_g & & \uparrow \mathrm{Ad}\, g & & \uparrow \mathrm{Ad}\, g \\ D(\rho) & \xrightarrow{\kappa_2} & \mathfrak{p}(\rho)^+ & \xrightarrow{\alpha_2} & \mathfrak{p}(\rho) \end{array}$$

<u>where</u> α_1 <u>and</u> α_2 <u>are linear isomorphisms defined by</u>

$$\alpha_1^{-1}(p) = \frac{1}{2}(p-\frac{i}{2}[j_W,p]) \quad \underline{\text{and}} \quad \alpha_2^{-1}(X) = \frac{1}{2}(X-\frac{i}{2}[j_\rho,X]) .$$

Proof. Let $Sp(W)^C$ be the subgroup of $GL(W^C)$ generated by $\mathfrak{sp}(W)^C$. Let $P(W)^{\pm}$, $U(W)^C$, $P(\rho)^{\pm}$, $U(\rho)^C$ be the subgroups of $Sp(W)^C$ generated by $\mathfrak{p}(W)^{\pm}$, $\mathfrak{u}(W)^C$, $\mathfrak{p}(\rho)^{\pm}$, $\mathfrak{u}(\rho)^C$, respectively. Then, as is well-known,

$$Sp(W) \subset P(\rho)^+U(\rho)^CP(\rho)^-$$

$$Sp(W) = gSp(W)g^{-1} \subset P(W)^+U(W)^CP(W)^- .$$

On the other hand we have $gP(\rho)^{\pm}g^{-1}=P(W)^{\pm}$, $gU(\rho)^Cg^{-1}=U(W)^C$. Let $x \in Sp(W)$. Then it can be written as $x=p^+kp^-$, where $p^{\pm} \in P(\rho)^{\pm}, k \in U(\rho)^C$. Consequently $gxg^{-1}=gp^+g^{-1} \cdot gkg^{-1} \cdot gp^-g^{-1}$, where $gp^{\pm}g^{-1} \in P(W)^{\pm}, gkg^{-1} \in U(W)^C$. There exists a unique $X \in \mathfrak{p}(\rho)^+$ such that $p^+=\exp X$. Hence, by the definition of κ_1, we have

$$\kappa_1 A_g(xU(\rho)) = \kappa_1(gxg^{-1}U(W)) = (\mathrm{Ad}\ g)X$$
$$= (\mathrm{Ad}\ g)\kappa_2(xU(\rho)) ,$$

which proves the commutativity of the left square. $\alpha_1 \mathrm{Ad}\ g = \mathrm{Ad}\ g \cdot \alpha_2$ is easily seen.

QED

Lemma 8.4. $\alpha_1\kappa_1(D(W))=\mathrm{Ad}\ g \cdot \alpha_2\kappa_2(D(\rho))$.

Proof. Suppose $gj_\rho g^{-1}=j_W$. Then the assertion is reduced to Lemma 8.3. Suppose $gj_\rho g^{-1}=-j_W$. Let $\bar{\kappa}_2$ be the Harish-Chandra imbedding of $D(\rho)$ into $\mathfrak{p}(\rho)^-$ and let $\bar{\alpha}_2$ be the linear isomorphism of $\mathfrak{p}(\rho)^-$ onto $\mathfrak{p}(\rho)$ defined by $\alpha_2^{-1}(X)=\frac{1}{2}(X+\frac{i}{2}[j_\rho,X])$. It can be verified that

$$\bar{\kappa}_2(D(\rho)) = \overline{\kappa_2(D(\rho))} ,$$

where the bar in the right-hand side is the conjugation of $\mathfrak{p}(\rho)^C$ with respect to $\mathfrak{p}(\rho)$. Let Ad^-g denote the map $\mathrm{Ad}\ g$ of $\mathfrak{p}(\rho)^-$ onto $\mathfrak{p}(W)^+$, in order to distinguish it from $\mathrm{Ad}\ g$ of $\mathfrak{p}(\rho)^+$ onto $\mathfrak{p}(W)^+$ in the case $gj_\rho g^{-1}=j_W$. Then $(\mathrm{Ad}^-g)\bar{X}=(\mathrm{Ad}\ g)X$ holds for $X \in \mathfrak{p}(\rho)^+$. So we have

$$\mathrm{Ad}^-g \cdot \bar{\kappa}_2(D(\rho)) = \mathrm{Ad}^-g \cdot \overline{\kappa_2(D(\rho))} = \mathrm{Ad}\ g \cdot \kappa_2(D(\rho))$$
$$= \kappa_1 A_g(D(\rho)) = \kappa_1(D(W)) ,$$

from which it follows that

$$\alpha_1\kappa_1(D(W)) = \alpha_1\mathrm{Ad}\,\bar{g}\cdot\bar{\kappa}_2(D(\rho)) = \mathrm{Ad}\ g\cdot\bar{\alpha}_2\bar{\kappa}_2(D(\rho))$$

$$= \mathrm{Ad}\ g\cdot\bar{\alpha}_2\overline{\kappa_2(D(\rho))} = \mathrm{Ad}\ g\cdot\alpha_2\kappa_2(D(\rho)) .$$

QED

Let us define an inner product $\langle\ ,\ \rangle$ on W as

$$\langle u,v\rangle = \rho(j_W u,v) .$$

Then each $p\in\mathfrak{p}(W)$ is a symmetric operator with respect to $\langle\ ,\ \rangle$.

Proposition 8.5.

$$\alpha_1\kappa_1(D(W)) = \{p\in\mathfrak{p}(W)\ ;\ \langle pu,pu\rangle < \langle u,u\rangle\ ,\ u\neq 0\ ,\ u\in W\} .$$

Proof. The subalgebra $\mathfrak{sp}(W)_u = \mathfrak{u}(\rho)+i\,\mathfrak{p}(\rho)$ is a compact real form of $\mathfrak{sp}(W)^C$. Let θ be the conjugation of $\mathfrak{sp}(W)^C$ with respect to $\mathfrak{sp}(W)_u$. Let us put $X^*=-\theta(X)$, $X\in\mathfrak{sp}(W)^C$. Then a lemma of Langlands [12] shows (see also Ise [6])

$$\kappa_2(D(\rho)) = \{X\in\mathfrak{p}(\rho)^+;\ 2-(\mathrm{ad}_{\mathfrak{p}(\rho)}-X^*)(\mathrm{ad}_{\mathfrak{p}(\rho)}-X) > 0\} . \qquad \ldots(**)$$

Let $\{e_1,\ldots,e_{2n}\}$ be the base of W chosen in the proof of Lemma 8.2. With respect to this base we have

$$\mathfrak{p}(\rho) = \left\{\begin{pmatrix} X_1 & X_2 \\ X_2 & -X_1\end{pmatrix}\ ;\ {}^tX_1=X_1\ ,\ {}^tX_2=X_2\right\} ,$$

and with respect to the base $\{e_1+ie_{n+1},\ldots, e_n+ie_{2n}, e_1-ie_{n+1},\ldots, e_n-ie_{2n}\}$ of W^C, we have

$$\mathfrak{p}(\rho)^+ = \left\{\begin{pmatrix} 0 & Z \\ 0 & 0\end{pmatrix}\ ;\ {}^tZ=Z\right\} ,$$

$$\mathfrak{p}(\rho)^- = \left\{\begin{pmatrix} 0 & 0 \\ Y & 0\end{pmatrix}\ ;\ {}^tY=Y\right\} .$$

Let $\begin{pmatrix} 0 & Z \\ 0 & 0\end{pmatrix} \in\mathfrak{p}(\rho)^+$. Then $\begin{pmatrix} 0 & Z \\ 0 & 0\end{pmatrix}^* = \begin{pmatrix} 0 & 0 \\ \bar{Z} & 0\end{pmatrix}\in\mathfrak{p}(\rho)^-$. Hence, for $\begin{pmatrix} 0 & 0 \\ Y & 0 \end{pmatrix} \in \mathfrak{p}(\rho)^-$, we have

$$\operatorname{ad} \begin{pmatrix} 0 & Z \\ 0 & 0 \end{pmatrix}^* \operatorname{ad} \begin{pmatrix} 0 & Z \\ 0 & 0 \end{pmatrix} \begin{pmatrix} 0 & 0 \\ Y & 0 \end{pmatrix} = \begin{pmatrix} 0 & 0 \\ \bar{Z}ZY+YZ\bar{Z} & 0 \end{pmatrix} .$$

From this and (**) it follows that

$$\kappa_2(D(\rho)) = \left\{ \begin{pmatrix} 0 & Z \\ 0 & 0 \end{pmatrix} \in \mathfrak{p}(\rho)^+ ; Z\bar{Z} < 1 \right\} .$$

So, by direct computations we can conclude

$$\alpha_2\kappa_2(D(\rho)) = \{X \in \mathfrak{p}(\rho) ; {}^tXX < 1\} .$$

Hence, from Lemma 8.4 it follows that

$$\alpha_1\kappa_1(D(W)) = \{gXg^{-1} \in \mathfrak{p}(W) ; {}^tXX < 1\} ,$$

which implies the lemma.

QED

We know in §4 that $\{\mathfrak{s}, \mathfrak{k}\}$ is a j-subalgebra of $\{\mathfrak{sp}(W), \mathfrak{u}(W)\}$. Let S^* and K^* be the subgroups of Sp(W) generated by $\mathfrak{s}$ and $\mathfrak{k}$, respectively. Then K^* is a maximal compact subgroup of S^* , and the classifying domain $D(\mathfrak{s})$ is represented as S^*/K^* .

<u>Lemma 8.6. The inclusions</u> $S^* \subset Sp(W)$ <u>and</u> $K^* \subset U(W)$ <u>induce the holomorphic imbedding</u> ϕ <u>of</u> $D(\mathfrak{s})$ <u>into</u> $D(W)$ <u>as the</u> S^*<u>-orbit of the origin</u>.

<u>Proof.</u> For each $s \in S^*$, ϕ is defined to be $\phi(sK^*)=sU(W)$. $\{\mathfrak{s}, \mathfrak{k}\}$ being a j-subalgebra, ϕ is holomorphic. Furthermore we have $\phi(S^*/K^*) = S^*/S^* \cap U(W)$, and the identity component of $S^* \cap U(W)$ is K^* . Hence ϕ is a holomorphic immersion. $S^* \cap U(W)$ is a compact subgroup of S^* containing K^* . So $S^* \cap U(W)=K^*$ and ϕ is an imbedding.

QED

$\{\mathfrak{s}, \mathfrak{k}\}$ being a j-subalgebra of $\{\mathfrak{sp}(W), \mathfrak{u}(W)\}$, we have $\mathfrak{p}^\pm \subset \mathfrak{p}(W)^\pm$.

<u>Lemma 8.7. Let</u> κ <u>be the Harish-Chandra imbedding of</u> $D(\mathfrak{s})$ <u>into</u> $\mathfrak{p}^+$,

and α be the restriction of α_1 to $\mathcal{P}^+$. Then

$$\alpha\kappa(D(\mathfrak{S})) \subset \{p \in \mathcal{P}\ ;\ \langle pu,pu\rangle < \langle u,u\rangle\ ,\ u \in W,\ u \neq 0\}\ .$$

Proof. As a special case of a result of Satake [16] we have the following commutative diagram

$$\begin{array}{ccccc}
D(W) & \xrightarrow{\kappa_1} & \mathcal{P}(W)^+ & \xrightarrow{\alpha_1} & \mathcal{P}(W) \\
\phi \uparrow & & \cup & & \cup \\
D(\mathfrak{S}) & \xrightarrow{\kappa} & \mathcal{P}^+ & \xrightarrow{\alpha} & \mathcal{P}
\end{array}\ .$$

Hence, from Lemma 8.5 we get the lemma.

QED

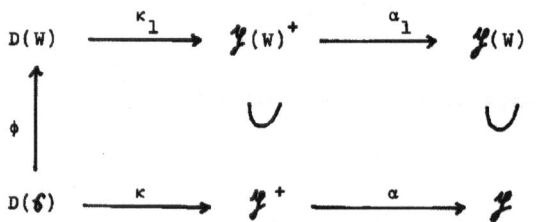

§9. Realizations as Siegel Domains of Type III

Let X_1 be a vector space over $\mathbb{R}$, X_1^C its complexification, and X_2 be a vector space over $\mathbb{C}$. A map L of $X_2 \times X_2$ to X_1^C is called a semi-hermitian form if L can be written as

$$L = L_1 + L_2 ,$$

where L_1 and L_2 are X_1^C-valued functions on $X_2 \times X_2$ satisfying the following conditions:

1) $L_1(u,v)$ is $\mathbb{C}$-linear in u.

2) $\overline{L_1(u,v)} = L_1(v,u)$, where the bar is the conjugation of X_1^C with respect to X_1.

3) $L_2(u,v)$ is a symmetric $\mathbb{C}$-bilinear form.

A semi-hermitian form L is called non-degenerate, if the condition "$L(u,v)=0$ for all $u \in X_2$" implies $v=0$.

Definition 9.1. Let X_1 and X_2 be the same as above. Let V be a convex cone in X_1 and B be a bounded domain in a vector space X_3 over $\mathbb{C}$. For each $p \in B$ let L_p be a non-degenerate semi-hermitian form of $X_2 \times X_2$ to X_1^C which depends differentiably on p. Then the open subset

$$\{(z,u,p) \in X_1^C \times X_2 \times X_3 \ ; \ \operatorname{Im} z - \operatorname{Re} L_p(u,u) \in V , \ p \in B\}$$

is called a Siegel domain of type III. In the definition the real part and the imaginary part are taken with respect to X_1.

Example. Let $X_1 = \mathbb{R}$, $V = \mathbb{R}^+$, $X_2 = X_3 = \mathbb{C}$, and $B = \{t \in X_3 ; |t|<1\}$. For each $t \in B$ let us put

$$L_t(u,v) = \frac{1}{1-|t|^2} (u\bar{v} + \bar{t}uv) .$$

Then $L_t(u,v)$ is a non-degenerate semi-hermitian form. The corresponding Siegel domain of type III is given as

$$\{(z,v,t) \in \mathbb{C}^3 \ ; \ \operatorname{Im} z - \frac{1}{1-|t|^2} \operatorname{Re}(|u|^2 + \bar{t}u^2) > 0, \ |t|<1\} ,$$

which is holomorphically equivalent to the symmetric bounded domain

$Sp(2,\mathbb{H})/U(2)$ (cf. [14]).

In what follows we will employ notations in all the preceding sections. Let D_1 be a homogeneous bounded domain and $D(V,F)\subset R^c\times W$ be the affine homogeneous Siegel domain of type II holomorphically equivalent to D_1. Let $\{\mathfrak{f}_1, j, \omega_1\}$ be the Iwasawa j-algebra of D_1 associated with the realization $D(V,F)$. Then $\mathfrak{f}_1$ can be written in the form $\mathfrak{f}_1 = jR+R+W$, where the subspaces in the right-hand side satisfy the bracket relations in Theorem 3.11. Let $\{\mathfrak{g}, \mathfrak{k}, j, \omega\}$ be the universal j-algebra of $\mathfrak{f}_1$: $\mathfrak{g}$ can be written as $\mathfrak{g} = \mathfrak{f}_1 + \mathfrak{s}$ (semi-direct), where $\mathfrak{s}$ is a semi-simple subalgebra of non-compact type (cf. §4). The pair $\{\mathfrak{s}, \mathfrak{k}\}$ is a j-subalgebra of $\{\mathfrak{g}, \mathfrak{k}\}$, to which there corresponds the symmetric bounded domain $D(\mathfrak{s})$, the classifying domain of D_1. We will always assume that <u>$D(\mathfrak{s})$ is non-trivial</u> (cf. Definition 5.9), and we identify $D(\mathfrak{s})$ with its image $\alpha\kappa(D(\mathfrak{s}))\subset \mathfrak{f}$ (cf. §8). By Lemma 8.7 the operator $1-p$ on W is non-singular for each $p\in D(\mathfrak{s})$.

<u>Definition 9.2.</u> For each $p\in D(\mathfrak{s})$ we define an R^c-valued function L_p on $W\times W$ as

$$L_p(u,v) = F(u,(1-p)^{-1}v) .$$

<u>Lemma 9.3.</u> <u>Let</u> $p\in \mathfrak{f}$. <u>Then, for an integer</u> $\nu\geq 0$, <u>we have</u>

1) $F(p^{2\nu}u,v) = F(u,p^{2\nu}v)$

2) $F(u, p^{2\nu+1}v) = F(v, p^{2\nu+1}u)$

3) $F(u, p^{2\nu}u) = F(p^{\nu}u, p^{\nu}u)$.

<u>Proof.</u> By Proposition 3.13 we can write F as

$$F(u,v) = \frac{1}{4}([j_W u,v] + i[u,v]) .$$

Using this we have

$$F(pu,v) = F(pv,u) ,$$

since $pj_W = -j_W p$ and $j_W, p\in \tilde{\mathfrak{s}}_1$. Using this equality repeatedly we have

$$F(p^{2\nu}u,v) = F(p(p^{2\nu-1}u),v) = F(pv, p^{2\nu-1}u)$$

$$= \overline{F(p(p^{2\nu-2}u),\ pv)} = \overline{F(p^2v,\ p^{2\nu-2}u)}$$

$$= F(p^{2\nu-2}u,\ p^2v) = \ldots\ldots = F(u,\ p^{2\nu}v)\ ,$$

which proves 1) .

2) and 3) are easily seen.

QED

<u>Lemma 9.4.</u> <u>For each</u> $p\varepsilon D(\mathfrak{f})$ <u>let us define the maps</u> $L_p^{(1)}$ <u>and</u> $L_p^{(2)}$ <u>as</u>

$$L_p^{(1)}(u,v) = F(u,(1-p^2)^{-1}v)\ ,$$

$$L_p^{(2)}(u,v) = F(u,(1-p^2)^{-1}pv)\ .$$

<u>Then</u> 1) $L_p^{(1)}(u,v)$ <u>is</u> $\mathbb{C}$<u>-linear in</u> u <u>and hermitian with respect to</u> R.

2) $L_p^{(2)}(u,v)$ <u>is</u> $\mathbb{C}$<u>-bilinear and symmetric</u>.

3) $L_p^{(1)}(u,u)\varepsilon\overline{V}$, $u\varepsilon W$.

<u>Proof.</u> 1) By Lemma 8.7, $p\varepsilon D(\mathfrak{f})$ is a symmetric operator on W with respect to $<\ ,\ >$, all eigenvalues of which have the absolute values smaller than one. Hence $(1-p^2)^{-1}$ can be represented as the convergent power series $\sum_{\nu=0}^{\infty} p^{2\nu}$. Using Lemma 9.3 we have

$$\overline{L_p^{(1)}(u,v)} = \overline{F(u,(1-p^2)^{-1}v)} = F((1-p^2)^{-1}v,u)$$

$$= F((\sum_{\nu=0}^{\infty} p^{2\nu})v,u) = \sum_{\nu=0}^{\infty} F(v,\ p^{2\nu}u)$$

$$= F(v,(1-p^2)^{-1}u) = L_p^{(1)}(v,u)\ ,$$

which proves 1).

2) Since the operator $(1-p^2)^{-1} = \sum_{\nu=0}^{\infty} p^{2\nu}$ is $\mathbb{C}$-linear, $L_p^{(2)}(u,v)$ is $\mathbb{C}$-bilinear. Furthermore

$$L_p^{(2)}(u,v) = F(u,(1-p^2)^{-1}pv) = \sum_{\nu=0}^{\infty} F(u,\ p^{2\nu+1}v)$$

$$= \sum_{\nu=0}^{\infty} F(v,\ p^{2\nu+1}u) = F(v,(1-p^2)^{-1}pu) = L_p^{(2)}(v,u)\ .$$

3) We have

$$L_p^{(1)}(u,u) = \sum_{\nu=0}^{\infty} F(u, p^{2\nu}u) = \sum_{\nu=0}^{\infty} F(p^{\nu}u, p^{\nu}u) .$$

Since $F(p^{\nu}u, p^{\nu}u)\in\overline{V}$ and $\overline{V}$ is a closed convex cone, we get

$$\sum_{\nu=0}^{\infty} F(p^{\nu}u, p^{\nu}u)\in\overline{V} .$$

QED

Corollary 9.5. The map L_p in Definition 9.2 is a non-degenerate semi-hermitian form.

Proof. Since $L_p = L_p^{(1)} + L_p^{(2)}$, L_p is a semi-hermitian form. Suppose $L_p(u,v)=0$ for all $u\in W$. Then putting $u=(1-p)^{-1}v$, we have

$$F((1-p)^{-1}v, (1-p)^{-1}v) = 0 ,$$

which implies $v=0$.

QED

By Corollary 9.5 we can consider the Siegel domain $\mathcal{S}$ of type III in $R^c \times W \times \mathcal{Y}$ defined by

$$\mathcal{S} = \{(z,v,p)\in R^c \times W \times \mathcal{Y} \ ; \ \mathrm{Im}\ z - \mathrm{Re}\ F(v,(1-p)^{-1}v)\in V \ , \ p\in D(\mathfrak{G})\} .$$

We will identify $\mathfrak{n} = R^c + W^+ + \mathcal{Y}^+$ with $R^c \times W \times \mathcal{Y}$ by the following linear isomorphism ψ ;

$$\psi(z,v,p) = z + \tfrac{1}{2}(v - ij_W v) + \tfrac{1}{2}(p - ij_W p) ,$$

where $z\in R^c$, $v\in W$, $p\in\mathcal{Y}$. Let T_1 be the Iwasawa group of D_1 generated by $\mathfrak{t}_1$, and c_1 be the Cayley transformation associated with D_1 (cf. §7). Then, since T_1 normalizes N by Lemma 7.1,8), it follows from Lemma 7.13 that

$$c_1 T_1 c_1^{-1}\cdot NL = c_1 T_1 NL = c_1 N T_1 L = N c_1 T_1 L \subset N N_1 L = NL ,$$

which implies that the group $c_1 T_1 c_1^{-1}$ leaves the orbit $N\cdot o$ stable. Hence $c_1 T_1 c_1^{-1}$ acts holomorphically on $R^c \times W \times \mathcal{Y}$ via ξ . We want to know this action explicitely. Since $T_1 = T_1' \cdot RW$ (cf. §7), we have $c_1 T_1 c_1^{-1} = c_1 T_1' c_1^{-1} \cdot RW$. So it is enough to determine the ξ-equivariant actions of $c_1 T_1' c_1^{-1}$ and of RW.

<u>Lemma 9.6.</u> <u>Let</u> $\mathfrak{g}$ <u>be a Lie algebra over</u> $\mathbb{R}$. <u>Suppose</u> $[\mathfrak{g},[\mathfrak{g},\mathfrak{g}]]=0$. <u>Then for</u> $X, Y \in \mathfrak{g}$ <u>we have</u>

1) $\exp X \exp Y = \exp((X+Y) + \frac{1}{2}[X,Y])$

2) $(\exp Y \exp X)^{-1} \exp X \exp Y = \exp [X,Y]$.

<u>Proof.</u> 1) is an immediate consequence of Lemma 4.1 of Ch. VI of Helgason [5].

2) is easily seen from 1).

QED

For an arbitrary $w \in W$ we write $w^{\pm}$ for $\frac{1}{2}(w \mp i j_W w)$. Then we have

<u>Lemma 9.7.</u> <u>Let</u> $X = \frac{1}{2}(p - i j_W p) \in \mathfrak{y}^+$, $p \in \mathfrak{y}$. <u>Then, for</u> $u, v \in W$, <u>we have</u>

1) $[X,u] = (pu)^+$

2) $\frac{1}{2}[u,[u,X]] = -iF(pu,u)$

3) $\frac{1}{2}[u^-, v^+] = iF(v,u)$.

The lemma can be derived from the bracket relations in $\mathfrak{s}_1$ and Proposition 3.13, 2).

<u>Proposition 9.8.</u> <u>The</u> <u>ξ-equivariant actions of the groups</u> RW <u>and</u> $c_1 T_1' c_1^{-1}$ <u>on</u> $R^c \times W \times \mathfrak{y}$ <u>are given by</u>:

1) $(\exp a \exp u)(z,v,p)$

$= (z+a+2iF(v,u) + iF(u,u) - iF(pu,u),\ v+u-pu, p)$

2) $(c_1 h c_1^{-1})(z,v,p) = ((Ad_R h)x + i(Ad_R h)y\ ,\ (Ad_W h)v, p)$

<u>where</u> $a \in R$, $u \in W$, $h \in T_1'$, <u>and</u> $z = x+iy \in R^c$, $v \in W$, $p \in \mathfrak{y}$.

<u>Proof.</u> 1) From Lemma 9.6 we get

$$\exp u = \exp u^+ \exp \tfrac{1}{2}[u^-,u^+] \exp u^-$$

$$\exp u^- \exp v^+ = \exp v^+ \exp u^- \exp [u^-,v^+] .$$

Considering the equalities, Lemma 7.1, 4), 6) and the bracket relations in $\mathfrak{f}_1$, we have, for $\psi(z,v,p)=z+v^++X$,

$$(\exp a \exp u)\exp(z+v^++X)L = \exp a \exp u \exp z \exp v^+ \exp X\cdot L$$

$$= \exp z \exp a \exp u \exp X(\exp u)^{-1} \exp u \exp v^+\cdot L$$

$$= \exp(z+a)\exp((\mathrm{Ad} \exp u)X)\exp u^+ \exp \tfrac{1}{2}[u^-,u^+] \exp u^- \exp v^+\cdot L$$

$$= \exp(z+a)\exp(X+ [u,X] +\tfrac{1}{2} [u, [u,X]])$$

$$\times \exp u^+ \exp \tfrac{1}{2} [u^-,u^+] \exp v^+ \exp u^- \exp [u^-,v^+] \cdot L$$

$$= \exp(z+a+X+ [u,X] +\tfrac{1}{2} [u, [u,X]] +u^++\tfrac{1}{2} [u^-,u^+] +v^++ [u^-,v^+])\cdot L .$$

Hence, considering Lemma 9.7, we get

$$\xi^{-1}(\exp a \exp u)\xi(z,v,p)$$

$$= (z+a+2iF(v,u)+iF(u,u)-iF(pu,u),\ v+u-pu,p) .$$

2) $c_1hc_1^{-1}\varepsilon L$ is valid, as is known in the proof of Lemma 7.12. It is seen that $[\mathrm{Ad}_W h, j_W]=0$. Hence we have

$$(c_1hc_1^{-1})\exp(z+v^++X)\cdot L = c_1hc_1^{-1} \exp z \exp v^+ \exp X\cdot L$$

$$= c_1hc_1^{-1} \exp(x+iy)(c_1hc_1^{-1})^{-1}(c_1hc_1^{-1})\exp v^+\cdot(c_1hc_1^{-1})^{-1}c_1hc_1^{-1} \exp X\cdot L$$

$$= \exp((\mathrm{Ad}\ c_1hc_1^{-1})(x+iy))\exp\left[\tfrac{1}{2}(\mathrm{Ad}\ c_1hc_1^{-1})(v-ij_Wv)\right] \exp X\cdot c_1hc_1^{-1}\cdot L$$

$$= \exp(\mathrm{Ad}c_1\ \mathrm{Ad}h\ \mathrm{Ad}c_1^{-1}(x+iy))\exp \left(\tfrac{1}{2}\mathrm{Ad}c_1\ \mathrm{Ad}h\ \mathrm{Ad}c_1^{-1}(v-ij_Wv)\right) \exp X\cdot L$$

$$= \exp((\mathrm{Ad}h)x+i(\mathrm{Ad}h)y)\exp(\tfrac{1}{2}((\mathrm{Ad}h)v-ij_W(\mathrm{Ad}h)v)\exp X\cdot L$$

$$= \exp((\mathrm{Ad}h)x+i(\mathrm{Ad}h)y+((\mathrm{Ad}h)v)^++X)\cdot L ,$$

where $z+v^++X=\psi(z,v,p)$. Hence we get

$$\xi^{-1}(c_1 h c_1^{-1})\xi(z,v,p) = ((Ad_R h)x+i(Ad_R h)y,\ (Ad_W h)v,p)$$

QED

<u>Lemma 9.9.</u> <u>The Siegel domain</u> $\mathcal{S}$ <u>of type</u> III <u>is stable under the</u> ξ-<u>equivariant actions of</u> RW <u>and</u> $c_1 T_1' c_1^{-1}$.

<u>Proof.</u> Let $(z,v,p)\in\mathcal{S}$. We put $(z',v',p')=(\exp a \exp u)(z,v,p)$. By Proposition 9.8 we have

$$\mathrm{Im} z'-\mathrm{Re}\, L_{p'}(v',v') = \mathrm{Im}(z+a+2iF(v,u)+iF(u,u)-iF(pu,u))$$

$$-\mathrm{Re}\, L_p(v+(1-p)u,\ v+(1-p)u)$$

$$= \mathrm{Im} z+2\mathrm{Re}\, F(v,u)+\mathrm{Re}\, F(u,u)-\mathrm{Re}\, F(pu,u)$$

$$-\mathrm{Re}\, L_p(v,v)-\mathrm{Re}\, L_p(v,(1-p)u)-\mathrm{Re}\, L_p((1-p)u,v)-\mathrm{Re}\, L_p((1-p)u,(1-p)u)$$

$$= \mathrm{Im} z+2\mathrm{Re}\, F(v,u)+F(u,u)-\mathrm{Re}\, F(pu,u)$$

$$-\mathrm{Re}\, L_p(v,v)-\mathrm{Re}\, F(v,u)-\mathrm{Re}\, L_p((1-p)u,v)-\mathrm{Re}\, F((1-p)u,u)$$

$$= \mathrm{Im} z-\mathrm{Re}\, L_p(v,v)+\mathrm{Re}(F(v,u)-L_p((1-p)u,v)) .$$

On the other hand we have

$$F(v,u)-L_p((1-p)u,v) = L_p(v,(1-p)u)-L_p((1-p)u,v)$$

$$= L_p^{(1)}(v,(1-p)u)-L_p^{(1)}((1-p)u,v) = 2i\ \mathrm{Im}\, L_p^{(1)}(v,(1-p)u)\in iR .$$

Hence $\mathrm{Re}(F(v,u)-L_p((1-p)u,v))=0$. Thus we get

$$\mathrm{Im} z'-\mathrm{Re}\, L_{p'}(v',v') = \mathrm{Im} z-\mathrm{Re}\, L_p(v,v) ,$$

which implies that RW leaves $\mathcal{S}$ stable.

Let us put $(z_1,v_1,p_1)=(c_1 h c_1^{-1})(z,v,p)$. Then by Proposition 9.8 we get

$$\mathrm{Im} z_1-\mathrm{Re}\, L_{p_1}(v_1,v_1) = (Ad_R h)y-\mathrm{Re}\, L_p((Ad_W h)v,\ (Ad_W h)v)$$

$$= (Ad_R h)y-\mathrm{Re}\, F((Ad_W h)v,\ (1-p)^{-1}(Ad_W h)v) .$$

The relation $[jR,\mathfrak{r}]=0$ implies $(1-p)^{-1}Ad_W h = Ad_W h\cdot(1-p)^{-1}$. So we have

$$F((Ad_W h)v,(1-p)^{-1}(Ad_W h)v) = (Ad_R h)F(v,(1-p)^{-1}v) = (Ad_R h)L_p(v,v) .$$

Hence we get

$$Im z_1 - Re\ L_{p_1}(v_1,v_1) = (Ad_R h)(y - Re\ L_p(v,v)) .$$

Since $Ad_R h$ leaves the cone V stable, the right-hand side is in V. Thus we proved that $c_1 T_1' c_1^{-1}$ leaves $\mathcal{S}$ stable.

QED

Let $(z,0,p)\varepsilon\mathcal{S}$ and let $\psi(z,0,p)=z+X$. Then $p\varepsilon\alpha\kappa(D(\mathfrak{r}))\subset\mathcal{Y}$ and $X\varepsilon\kappa(D(\mathfrak{r}))$. Since the group S acts on $\kappa D(\mathfrak{r})\subset\mathcal{Y}^+$ equivariantly, for $s\varepsilon S$ we have

$$s\ \exp X\cdot K^C P^- = (\exp sX)K^C P^- .$$

Hence

$$s\ \exp X\cdot L = \exp sX\cdot L .$$

<u>Lemma 9.10.</u> S <u>acts on</u> $\xi^{-1}c_1\beta(D)$ <u>ξ-equivariantly. Furthermore let</u> $(z,0,p)\varepsilon\mathcal{S}$ <u>and</u> $s\varepsilon S$. <u>Then</u>

$$s\cdot(z,0,p) = (z,0,s\cdot p) .$$

<u>Proof.</u> We have

$$(\xi^{-1}S\xi)(\xi^{-1}c_1\beta(D)) = \xi^{-1}Sc_1\beta(D) = \xi^{-1}c_1 S\beta(D)$$

$$= \xi^{-1}c_1\beta(S\cdot D) = \xi^{-1}c_1\beta(D) ,$$

which shows the first assertion. With notations above

$$s\xi(z+X) = s\ \exp(z+X)\cdot L = s\ \exp z\ \exp X\cdot L$$

$$= \exp z\cdot s\ \exp X\cdot L = \exp z\ \exp sX\cdot L = \exp(z+sX)\cdot L .$$

So we get

$$s\cdot(z,0,p) = \xi^{-1}s\xi(z+X) = (z,0,s\cdot p) .$$

QED

Thus we obtain the main theorem:

Theorem 9.11. *Let* D *be the universal domain of a homogeneous bounded domain* D_1 *. Suppose that* D_1 *has the non-trivial classifying domain* $D(\mathfrak{S})$ (cf. Def. 5.9). *Let* $D(V,F) \subset R^c \times W$ *be the affine homogeneous Siegel domain of type* II *holomorphically equivalent to* D_1 *. Let* $\mathcal{S}$ *be the Siegel domain of type* III *defined by*

$$\{(z,v,p) \varepsilon R^c \times W \times \mathfrak{F} \; ; \; \mathrm{Im} z - \mathrm{Re}\, F(v,(1-p)^{-1}v) \varepsilon V, \; p \varepsilon D(\mathfrak{S})\} \; ,$$

where we identify $D(\mathfrak{S})$ *with its image* $\alpha\kappa(D(\mathfrak{S})) \subset \mathfrak{F}$. *Then the Cayley transform* $\xi^{-1} c_1 \beta(D)$ *of* D *coincides with* $\mathcal{S}$.

Proof. We know that

$$G_h = T_1 P K_h$$

$$P \subset P^+ K^C P^- \; .$$

Hence an arbitrary element $g \varepsilon G_h$ can be written as

$$g = \exp a \exp u \cdot hxk$$

with $a \varepsilon R$, $u \varepsilon W$, $h \varepsilon T_1'$, $x \varepsilon P$, $k \varepsilon K_h$; we can write x as $x = p^+ k_1 p^-$, $p^\pm \varepsilon P^\pm$, $k_1 \varepsilon K^C$, and there exists a unique $X \varepsilon \mathfrak{F}^+$ such that $p^+ = \exp X$. Hence

$$g = \exp a \exp u \cdot h \exp X \cdot k_1 p^- k \; .$$

Using Lemma 7.1, 6), bracket relations in $\mathfrak{f}_1$ and an equality in the proof of Lemma 7.13, we have

$$c_1 gL = c_1 \exp a \exp u \cdot h \exp X \cdot L = c_1 \exp a \exp u \exp X \cdot hL$$

$$= c_1 \exp a \exp u \exp X (\exp u)^{-1} \exp u \cdot hL$$

$$= c_1 \exp a \exp(X + [u,X] + \tfrac{1}{2} [u, [u,X]]) \exp u \cdot hL$$

$$= \exp(X + [u,X] + \tfrac{1}{2} [u, [u,X]]) c_1 \exp a \exp u \cdot hL$$

$$= \exp(X + [u,X] + \tfrac{1}{2} [u, [u,X]]) \exp(i(\mathrm{Ad}h)r + a + \tfrac{1}{2} [u^-, u^+] + u^+) L$$

$$= \exp(i(\mathrm{Ad}h)r + a + \tfrac{1}{2} [u^-, u^+] + \tfrac{1}{2} [u, [u,X]] + u^+ + [u,X] + X) L \; .$$

So, putting $\psi^{-1}(X)=p$, we have

$$\xi^{-1}c_1(g\cdot o) = (i(Adh)r+a+iF(u,u)-iF(pu,u),\ u-pu,p)\ .$$

Furthermore we have

$$Im(i(Adh)r+a+iF(u,u)-iF(pu,u))-Re\ L_p(u-pu,\ u-pu)$$

$$= (Adh)r+F(u,u)-Re\ F(pu,u)-Re\ F(u-pu,u) = (Adh)r\epsilon V,$$

which proves $\xi^{-1}c_1\beta(D) \subset \mathcal{S}$.

We remark that the group $c_1Gc_1^{-1}=c_1T_1c_1^{-1}\cdot S$ acts on $\xi^{-1}c_1\beta(D)$ ξ-equivariantly. Let $(z,v,p)\epsilon\mathcal{S}$. Since $(ir,0,0)\epsilon\xi^{-1}c_1\beta(D)$, in order to prove $\xi^{-1}c_1\beta(D)\supset\mathcal{S}$, it is enough to find an element of $c_1Gc_1^{-1}$ which carries (z,v,p) to $(ir,0,0)$. Put $u=-(1-p)^{-1}v$ and put $(\exp u)(z,v,p)=(z',v',p')$. Then by Proposition 9.8 we have $v'=v+u-pu=0$ and $p'=p$. Put $a=-Rez'$. Then $(\exp a)(z',v',p')=(iy',0,p)$, where $y'=Imz'$. Since $\exp a \exp u\ \epsilon\ c_1T_1c_1^{-1}$, the point $(iy',0,p)$ is in $\mathcal{S}$ by Lemma 9.9. So we have

$$y' = Im(iy')-ReL_p(0,0)\epsilon V\ .$$

Hence there exists $h\epsilon T_1'$ such that $(Adh)y'=r$. By Lemma 9.9 we get

$$c_1hc_1^{-1}(iy',0,p) = (ir,0,p)\ .$$

Since $0\epsilon D(\mathcal{V})$, we can find $s\epsilon S$ such that $s\cdot p=0$. By Lemma 9.10 we have

$$s\cdot(ir,0,p) = (ir,0,0)\ .$$

Furthermore $s\cdot c_1hc_1^{-1}\cdot\exp a \exp u\ \epsilon\ S\cdot c_1T_1c_1^{-1}=c_1T_1c_1^{-1}\cdot S = c_1Gc_1^{-1}$, which completes the proof.

QED

As an application of Theorem 9.11 we will reproduce the following theorem of Piatetski-Sapiro [15] .

<u>Theorem 9.12. Let D_o be a homogeneous bounded domain, $\{\mathfrak{g}_o,j,\omega\}$ be its Iwasawa j-algebra and $\mathfrak{g}_o=\mathfrak{g}_1+\mathfrak{g}_2$ be the decomposition given by (#) in §4 such that $\mathfrak{g}_1$ is a j-ideal and $\mathfrak{g}_2$ is a j-subalgebra. Let D_1</u>

and D_2 *be the homogeneous bounded domains corresponding to* $\mathfrak{f}_1$ *and* $\mathfrak{f}_2$ *, respectively. Let* $D(V,F) \subset R^c \times W$ *be the affine homogeneous Siegel domain of type II holomorphically equivalent to* D_1 *. Let* $D(\mathfrak{S})$ *be the classifying domain of* D_1 *and* μ *be the classifying map of* D_2 *into* $D(\mathfrak{S})$ (cf. §5). *Then* D_0 *is holomorphically equivalent to the Siegel domain of type III*

$$\mathcal{S}_0 = \{(z,v,t)\varepsilon R^c \times W \times \mathbb{C}^r \ : \ \mathrm{Im} z - \mathrm{Re}\, F(v,(1-\eta\mu(t))^{-1}v)\varepsilon V, t\varepsilon D_2\} \ ,$$

where $\eta = \alpha\kappa$ (cf. §8) *and* $D_2 \subset \mathbb{C}^r$, $\dim D_2 = r$.

Proof. From §5 and the proof of Theorem 9.11 we get the following commutative diagram:

$$\begin{array}{ccccc} D_0 & \xrightarrow{\ \tilde{\mu}\ } & D & \xrightarrow{\ \tilde{\eta}\ } & \mathcal{S} \\ \downarrow{\scriptstyle \pi_0} & & \downarrow{\scriptstyle \pi} & & \downarrow{\scriptstyle \pi_{\mathcal{S}}} \\ D_2 & \xrightarrow{\ \mu\ } & D(\mathfrak{S}) & \xrightarrow{\ \eta\ } & D(\mathfrak{S}) \subset \mathcal{Y} \end{array}$$

where $\tilde{\eta} = \xi^{-1} c_1 \beta$, $\pi_{\mathcal{S}}(z,v,p) = p$. We define the holomorphic map Φ of D_0 into $\mathcal{S} \times D_2$ as

$$\Phi(x) = (\tilde{\eta}\tilde{\mu}(x), \ \pi_0(x)) \ .$$

Φ is injective; in fact, let $x, x' \varepsilon D_0$ and suppose $(\tilde{\eta}\tilde{\mu}(x), \pi_0(x)) = (\tilde{\eta}\tilde{\mu}(x'), \pi_0(x'))$. Then $\pi_0(x) = \pi_0(x')$ and $\tilde{\mu}(x) = \tilde{\mu}(x')$. Since $\tilde{\mu}$ is diffeomorphic on each fibre, we get $x = x'$. Let us denote by Φ_*, π_{0*}, $\tilde{\eta}_*$, $\tilde{\mu}_*$ the differentials of the corresponding maps. Let $X \varepsilon T_x(D_0)$ and suppose $\Phi_*(X) = 0$. Then $\pi_{0*}(X) + \tilde{\eta}_*\tilde{\mu}_*(X) = 0$. Hence $\pi_{0*}(X) = \tilde{\mu}_*(X) = 0$. $\pi_{0*}(X) = 0$ implies that X is tangent to the fibre. So $\tilde{\mu}_*(X) = 0$ implies $X = 0$. Φ is thus a holomorphic imbedding of D_0 into $\mathcal{S} \times D_2$. We will show

$$\Phi(D_0) = \{(s,y)\varepsilon \mathcal{S} \times D_2 \ ; \ \eta\mu(y) = \pi_{\mathcal{S}}(s)\} \ . \qquad \ldots(*)$$

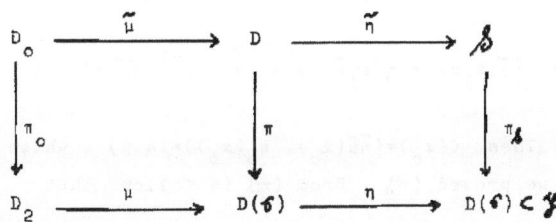

The commutativity of the diagram implies that $\Phi(D_0)$ is in the set of the right-hand side. Let $(s,y)\in\mathcal{S}\times D_2$ and suppose $\eta\mu(y)=\pi_{\mathcal{S}}(s)$. To the natural action of T_1 on D there corresponds the action of $c_1T_1c_1^{-1}$ on $\mathcal{S}$; in fact, for $t_1\in T_1$ and $x\in D$ we have

$$\tilde{\eta}(t_1x) = \xi^{-1}c_1\beta(t_1x) = \xi^{-1}c_1t_1\beta(x)$$

$$= \xi^{-1}c_1t_1c_1^{-1}c_1\beta(x) = (c_1t_1c_1^{-1})\xi^{-1}c_1\beta(x) .$$

Now take a point $x\in\pi_o^{-1}(y)$. Then

$$\pi_{\mathcal{S}}\tilde{\eta}\tilde{\mu}(x) = \eta\mu\pi_o(x) = \eta\mu(y) = \pi_{\mathcal{S}}(s) ,$$

which implies that $\tilde{\eta}\tilde{\mu}(x)\in\pi_{\mathcal{S}}^{-1}(\pi_{\mathcal{S}}(s))$. Since $c_1T_1c_1^{-1}$ acts transitively on each fibre of $\mathcal{S}$, we can find $t_1\in T_1$ such that $s=(c_1t_1c_1^{-1})\tilde{\eta}\tilde{\mu}(x)$. We have

$$\tilde{\eta}\tilde{\mu}(t_1x) = \tilde{\eta}(t_1\tilde{\mu}(x)) = (c_1t_1c_1^{-1})\tilde{\eta}\tilde{\mu}(x) = s .$$

Put $t_1x=x_o$. Then $\Phi(x_o)=(\tilde{\eta}\tilde{\mu}(x_o),\ \pi_o(x_o))=(s,y)$, which implies $(s,y)\in\Phi(D_o)$. Thus we proved (*). From (*) it follows that

$$\Phi(D_o) = \{((z,v,p),y)\in R^c\times W\times\mathcal{J}\times\mathbb{C}^r \ ;\ (z,v,p)\in\mathcal{S},\ y\in D_2,\ p=\eta\mu(y)\} ,$$

which is naturally identified with

$$\{(z,v,y)\in R^c\times W\times\mathbb{C}^r \ ;\ \mathrm{Im}\,z-\mathrm{Re}\,F(v,(1-\eta\mu(y))^{-1}v)\in V,\ y\in D_2\}.$$

Hence Φ is regarded as a holomorphic imbedding of D_o into $R^c\times W\times\mathbb{C}^r$. The image $\Phi(D_o)$ is a domain $\mathcal{S}_o$ in $R^c\times W\times\mathbb{C}^r$ and D_o is holomorphically equivalent to $\mathcal{S}_o$.

QED

Appendix

We will give here in the appendix an outline of the proof of the fundamental theorem of Vinberg, Gindikin and Piatetski-Sapiro [20], mentioned in §2;

Theorem A1. ([20]) Every homogeneous bounded domain in $\mathbb{C}^n$ is holomorphically equivalent to an affine homogeneous Siegel domain of type II or type I .

Let $\{\mathfrak{g}, \mathfrak{k}, (j), \omega\}$ be an effective proper j-algebra (cf. Def. 4.2). Suppose $\mathfrak{g}$ is semi-simple. Then by Borel [1] $\mathfrak{k}$ is a maximal compact subalgebra of $\mathfrak{g}$, and to the j-algebra $\{\mathfrak{g}, \mathfrak{k}, (j), \omega\}$ there corresponds a symmetric bounded domain.

Suppose that $\mathfrak{g}$ is not semi-simple. Then it is known in [20] that there exists a non-zero abelian ideal R of $\mathfrak{g}$ such that there exists an element $r \in R$ satisfying $[ja, r] = a$ for all $a \in R$. r is uniquely determined and is called the unit of R . An abelian ideal with such r is called of the first kind. There exists a $j \in (j)$ satisfying

$$[jr, \mathfrak{k}] = 0$$

(cf. [20]). Consider the Jordan decomposition of the complexification $\mathfrak{g}^C$ of $\mathfrak{g}$ by the operator $\mathrm{ad}\, jr$. Let $\tilde{\mathfrak{g}}^\lambda$ be the direct sum of all the Jordan eigenspaces of $\mathrm{ad}\, jr$ on which the corresponding eigenvalues have λ as their real parts. Then we have

$$\mathfrak{g}^C = \sum_\lambda \tilde{\mathfrak{g}}^\lambda \quad \text{(vector space direct sum)} .$$

Each $\tilde{\mathfrak{g}}^\lambda$ is stable under the conjugation of $\mathfrak{g}^C$ with respect to $\mathfrak{g}$. Hence we have

$$\mathfrak{g} = \sum_\lambda \mathfrak{g}^\lambda \qquad \ldots (*)$$

where $\mathfrak{g}^\lambda = \tilde{\mathfrak{g}}^\lambda \cap \mathfrak{g}$. There exists an $\mathrm{ad}_{\mathfrak{g}} \mathfrak{k}$ -invariant inner product on $\mathfrak{g}$ (cf. [20]) : each operator of $\mathrm{ad}_{\mathfrak{g}} \mathfrak{k}$ is skew-symmetric relative to the inner product. Hence the decomposition (*) is independent of the choice of $j \in (j)$. If $\mathfrak{g}^\lambda \neq (0)$, then λ is equal to one of 0 , 1 and $\frac{1}{2}$ (cf. [20]). We have thus the decomposition

$$\mathfrak{g} = \mathfrak{g}^0 + \mathfrak{g}^1 + \mathfrak{g}^{\frac{1}{2}} \qquad (**)$$

and $\mathfrak{g}$ is a graded Lie algebra, namely

$$[\mathfrak{g}^\lambda, \mathfrak{g}^\mu] \subset \mathfrak{g}^{\lambda+\mu}$$

holds. Moreover we have

$$j\,\mathfrak{g}^1 \subset \mathfrak{g}^0 \qquad \text{for all} \quad j \in (j) ,$$

$$j\,\mathfrak{g}^{\frac{1}{2}} = \mathfrak{g}^{\frac{1}{2}} \qquad \text{for some} \quad j \in (j) .$$

(cf. [20]). We need the following structure theorem of effective proper j-algebras.

Theorem A2. ([20]) Let $\{\mathfrak{g}, \mathfrak{k}, (j), \omega\}$ be an effective proper j-algebra and let R be a maximal abelian ideal of the first kind and r be its unit. Then there exists a j-ideal $\mathfrak{g}_1$ and a j-subalgebra $\mathfrak{s}_1$ which is semi-simple and of non-compact type, such that $\mathfrak{g}$ is the semi-direct sum of $\mathfrak{g}_1$ and $\mathfrak{s}_1$. Furthermore in the decomposition (**) by adjr we have

1) $\mathfrak{g}_1 = \mathfrak{g}_1 \cap \mathfrak{g}^0 + \mathfrak{g}^1 + \mathfrak{g}^{\frac{1}{2}}$,

2) $\mathfrak{g}^0 = \mathfrak{s}_1 + \mathfrak{g}_1 \cap \mathfrak{g}^0$,

3) $\mathfrak{g}_1 \cap \mathfrak{g}^0 = \mathfrak{g}_1 \cap \mathfrak{k} + jR$ for j leaving $\mathfrak{g}_1$ stable,

4) $\mathfrak{k} = \mathfrak{s}_1 \cap \mathfrak{k} + \mathfrak{g}_1 \cap \mathfrak{k}$,

5) $[\mathfrak{s}_1, \mathfrak{g}^1] = 0$, $\mathfrak{g}^1 = R$,

where the right-hand sides of 1) - 4) are vector space direct sums.

In the theorem above, if $\mathfrak{g}$ is semi-simple, then $\mathfrak{g} = \mathfrak{s}_1$. There may occur the case where $\mathfrak{g}^{\frac{1}{2}} = (0)$. We remark that if a j-algebra $\{\mathfrak{g}, \mathfrak{k}, (j), \omega\}$ is effective, then $\mathfrak{g}$ is centerless (cf. [20]), and consequently $\mathfrak{g}$ can be assumed to be a linear Lie algebra.

Theorem A3. ([20]) Suppose $\mathfrak{s}_1 = (0)$ in Theorem A2. Then there exists a Siegel domain $\mathfrak{D}$ of type II or type I in the vector space $\mathfrak{g}^1 + i\,\mathfrak{g}^1 + \mathfrak{g}^{\frac{1}{2}}$ such that the connected Lie group G without center, cor-

<u>responding to $\mathfrak{g}$, acts on $\mathfrak{D}$ as affine automorphisms; in that case $\mathfrak{D}$ can be represented as</u> G/K , <u>where</u> K <u>is the subgroup generated by $\mathfrak{k}$. If $\mathfrak{g}^{\frac{1}{2}}=(0)$, then $\mathfrak{D}$ is of type I</u>.

A j-algebra $\{\mathfrak{g}, \mathfrak{k}, (j), \omega\}$ is called <u>linear</u>, if $\mathfrak{g}$ is a linear Lie algebra. By the <u>algebraic hull</u> of a linear Lie algebra we mean the smallest algebraic subalgebra containing it.

<u>Theorem A4.</u> ([20]) <u>Let</u> $\{\mathfrak{g}, \mathfrak{k}, (j), \omega\}$ <u>be an effective linear j-algebra</u>. <u>Then there exists a linear j-algebra</u> $\{\hat{\mathfrak{g}}, \hat{\mathfrak{k}}, (\hat{j}), \hat{\omega}\}$ <u>satisfying the following conditions</u>:

1) <u>$\hat{\mathfrak{g}}$ is the algebraic hull of $\mathfrak{g}$ and $\hat{\mathfrak{k}}$ is an algebraic subalgebra</u>.

2) $\hat{\mathfrak{g}} = \mathfrak{g} + \hat{\mathfrak{k}}$ <u>and</u> $\mathfrak{g} \wedge \hat{\mathfrak{k}} = \mathfrak{k}$.

3) $\{\mathfrak{g}, \mathfrak{k}, (j), \omega\}$ <u>is a j-subalgebra of</u> $\{\hat{\mathfrak{g}}, \hat{\mathfrak{k}}, (\hat{j}), \hat{\omega}\}$. <u>Furthermore if $\mathfrak{g}$ is proper, then $\hat{\mathfrak{g}}$ is proper</u>.

We remark here that $\{\hat{\mathfrak{g}}, \hat{\mathfrak{k}}, (\hat{j}), \hat{\omega}\}$ can be assumed to be effective (cf. Lemma 3.2 in [7]).

<u>The Proof of Theorem A1.</u> Let D be a homogeneous bounded domain in $\mathbb{C}^n$. Let G be the identity component of the full automorphism group of D and K be the isotropy subgroup at a point of D . The pair $\{\mathfrak{g}, \mathfrak{k}\}$ of the Lie algebras is an effective proper (linear) j-algebra (cf. Lemma 4.3). Let $\{\hat{\mathfrak{g}}, \hat{\mathfrak{k}}\}$ be the j-algebra in Theorem A4 corresponding to the j-algebra $\{\mathfrak{g}, \mathfrak{k}\}$. Let $\tilde{D}$ be the universal covering manifold of D . Then there exists a connected Lie group $\hat{G}$ corresponding to $\hat{\mathfrak{g}}$ such that $\hat{G}$ acts on $\tilde{D}$ effectively, holomorphically and transitively (cf. §6). $\tilde{D}$ can be written as $\tilde{D}=\hat{G}/\hat{K}$, where $\hat{K}$ is the subgroup of $\hat{G}$ generated by $\hat{\mathfrak{k}}$.

Let G^* be a connected Lie subgroup of $\hat{G}$ of the smallest dimension with such properties that it acts on $\tilde{D}$ transitively and that its Lie algebra is an algebraic subalgebra of $\hat{\mathfrak{g}}$. We can assume that G^* is not semi-simple. We can write $\tilde{D}$ as $\tilde{D}=G^*/K^*$, where K^* is the isotropy subgroup at the origin o of $\tilde{D}=\hat{G}/\hat{K}$. The pair $\{\mathfrak{g}^*, \mathfrak{k}^*\}$ of their Lie algebras is an effective proper j-algebra. Hence, by Theorem A2 we can write $\mathfrak{g}^*$ as

$$\mathfrak{g}^* = \mathfrak{s}_1 + \mathfrak{g}_1 .$$

<u>Lemma 1.</u> $\mathfrak{g}_1$ <u>is an algebraic subalgebra</u>.

<u>Proof.</u> Let $\hat{\mathfrak{g}}_1$ be the algebraic hull of $\mathfrak{g}_1$. Then $\hat{\mathfrak{g}}_1$ is an ideal of $\mathfrak{g}^*$, since $\mathfrak{g}_1$ is an ideal. We have

$$\hat{\mathfrak{g}}_1 = \mathfrak{s}_1 \cap \hat{\mathfrak{g}}_1 + \mathfrak{g}_1 .$$

Putting $\hat{\mathfrak{s}}_1 = \hat{\mathfrak{g}}_1 \cap \mathfrak{s}_1$, $\hat{\mathfrak{s}}_1$ is an ideal of $\mathfrak{s}_1$ and so $\hat{\mathfrak{s}}_1$ is semi-simple. Hence

$$\hat{\mathfrak{s}}_1 = [\hat{\mathfrak{s}}_1, \hat{\mathfrak{s}}_1] \subset [\hat{\mathfrak{g}}_1, \hat{\mathfrak{g}}_1] \subset \mathfrak{g}_1$$

which implies $\hat{\mathfrak{s}}_1 = (0)$. So we get $\hat{\mathfrak{g}}_1 = \mathfrak{g}_1$.

QED

We will continue the proof of Theorem A1. Let S_1 and G_1 be the subgroups of G^* generated by $\mathfrak{s}_1$ and $\mathfrak{g}_1$, respectively. G_1 is a closed normal subgroup by Lemma 1. S_1 is also closed. Hence G^* can be written as

$$G^* = S_1 G_1 .$$

Let K_s be the subgroup of S_1 generated by $\mathfrak{k}_s = \mathfrak{s}_1 \cap \mathfrak{k}^*$. By using the fact that $\hat{\mathfrak{k}}$ is algebraic, it can be proved that $\hat{K}$ is a compact subgroup of $\hat{G}$. Hence $\mathfrak{k}_s$ is a compact subalgebra of $\mathfrak{s}_1$, and so the j-algebra $\{\mathfrak{s}_1, \mathfrak{k}_s\}$ is effective. Furthermore it is proper and $\mathfrak{s}_1$ is semi-simple, of non-compact type. Therefore Borel's Theorem [1] implies that $\mathfrak{k}_s$ is a maximal compact subalgebra of $\mathfrak{s}_1$. So K_s is a maximal compact subgroup of S_1. Let $S_1 = K_s \cdot T_s$ be the usual Iwasawa decomposition of S_1, T_s being the Iwasawa subgroup. Here we need

<u>Lemma 2.</u> <u>The subgroup</u> $T_s G_1$ <u>acts on</u> $\tilde{D}$ <u>transitively</u>.

<u>Proof.</u> We can prove by the similar method as in Lemma 5.3 that $\tilde{D}$ is a homogeneous Kähler manifold. Furthermore we have

$$\mathfrak{g}^* = \mathfrak{k}^* + (\mathfrak{t}_s + \mathfrak{g}_1)$$

$$\mathfrak{k}^* \cap (\mathfrak{t}_s + \mathfrak{g}_1) = \mathfrak{k}_1$$

where $\mathfrak{k}_1 = \mathfrak{k}^* \cap \mathfrak{g}_1$ and $\mathfrak{t}_s$ is the Lie algebra of T_s. Hence the orbit of T_sG_1 through o in $\tilde{D}$ is open in $\tilde{D}$, and consequently it coincides with $\tilde{D}$ itself (cf. Proposition 5.4).

QED

On the other hand the Lie algebras $\mathfrak{t}_s$ and $\mathfrak{g}_1$ are algebraic and so $\mathfrak{t}_s + \mathfrak{g}_1$ is algebraic. Hence, by the definition of $\mathfrak{g}^*$ we get

$$\mathfrak{g}^* = \mathfrak{t}_s + \mathfrak{g}_1$$

which implies $\mathfrak{k}_s = (0)$ and consequently $\mathfrak{r}_1 = (0)$. So we conclude $\mathfrak{g}^* = \mathfrak{g}_1$. By Theorem A3 there exists an affine homogeneous Siegel domain $\mathcal{D}$ on which a connected Lie group G_1 without center, corresponding to $\mathfrak{g}_1$, acts transitively, and $\mathcal{D}$ can be written as

$$\mathcal{D} = G_1/K_1 ,$$

where K_1 is the analytic subgroup generated by $\mathfrak{k}_1 = \mathfrak{k}^*$. Since G_1 is centerless, there exists the covering map ϕ of G^* onto G_1 which carries K^* to K_1. Hence ϕ induces the (holomorphic) covering map ϕ_1 of $\tilde{D} = G^*/K^*$ onto $\mathcal{D} = G_1/K_1$. On the other hand $\mathcal{D}$ is a cell and so ϕ_1 is a holomorphic homeomorphism. Thus we proved that the universal covering manifold $\tilde{D}$ of the homogeneous bounded domain D is a Siegel domain $\mathcal{D}$. Hence by Proposition 6.3 D itself is holomorphically equivalent to the Siegel domain $\mathcal{D}$. The proof of Theorem A1 is now completed.

QED

Bibliography

1 A. Borel, Kählerian coset spaces of semi-simple Lie groups, Proc. Nat. Acad. Sci. U.S.A. 40 (1954), 1147-1151.

2 A. Borel and Harish-Chandra, Arithmetic subgroups of algebraic groups, Ann. of Math. 75 (1962), 485-535.

3 A. Borel and R. Remmert, Über kompakte homogene Kählersche Mannigfaltigkeiten, Math. Ann. 145 (1962), 429-439.

4 C. Chevalley, Théorie des groupes de Lie, Hermann, Paris, 1968.

5 S. Helgason, Differential geometry and symmetric spaces, Academic Press, New York, 1962.

6 M. Ise, On canonical realization of bounded symmetric domains as matrix spaces, to appear.

7 S. Kaneyuki, On the automorphism groups of homogeneous bounded domains, J. Fac. Sci. Univ. Tokyo, 14 (1967), 89-130.

8 S. Kaneyuki and M. Sudo, On Šilov boundaries of Siegel domains, J. Fac. Sci. Univ. Tokyo, 15 (1968), 131-146.

9 S. Kobayashi, Invariant distances on complex manifolds and holomorphic mappings, J. Math. Soc. Japan, 19 (1967), 460-480.

10 S. Kobayashi and K. Nomizu, Foundation of differential geometry, Vol. I, Interscience, New York, 1963.

11 J.L. Koszul, Sur la forme hermitienne canonique des espaces homogènes complexes, Canad. J. Math. 7 (1955), 562-576.

12 R.P. Langlands, The dimension of spaces of automorphic forms, Amer. Jour. Math. 85 (1963), 99-125.

13 Y. Matsushima, On hermitian symmetric spaces, Summer Seminar in Diff. Geometry, Akakura, Japan, 1956 (in Japanese).

14 I.I. Piatetski-Sapiro, Geometry of classical domains and theory of automorphic functions, Fizmatgiz, Moscow, 1961. French translation, Paris, 1966.

15 I.I. Piatetski-Sapiro, Geometry and classification of homogeneous

bounded domains, Uspehi Math. Nauk, 20 (1965), 3-51. Russian Math. Surv. 20 (1966), 1-48.

16 I. Satake, A note on holomorphic imbeddings and compactification of symmetric domains, Amer. Jour. Math. 90 (1968), 231-247.

17 N. Tanaka, On infinitesimal automorphisms of Siegel domains, J. Math. Soc. Japan 22 (1970), 180-212.

18 E.B. Vinberg, Theory of homogeneous convex cones, Trudy Moskva Math. Obšč. 12(1963), 303-358. Trans. Moscow Math. Soc. (1963), 340-403.

19 E.B. Vinberg, Morozov-Borel theorem for real Lie groups, Dokl. Acad. Nauk U.S.S.R. 141 (1961), 270-273. Amer. Math. Soc. Trans. Vol. 2, No. 6 (1961).

20 E.B. Vinberg, S.G. Gindikin and I.I. Piatetski-Sapiro, On classification and canonical realization of complex homogeneous bounded domains, Trudy Moskva Math. Obšč. 12 (1963), 359-388. Trans. Moscow Math. Soc. 12 (1963), 404-437.

21 J.A. Wolf and A. Korányi, Generalized Cayley transforms of bounded symmetric domains, Amer. Jour. Math. 87 (1965), 899-939.

Lecture Notes in Mathematics

Comprehensive leaflet on request

Vol. 38: R. Berger, R. Kiehl, E. Kunz und H.-J. Nastold, Differentialrechnung in der analytischen Geometrie IV, 134 Seiten. 1967 DM 12,-

Vol. 39: Séminaire de Probabilités I. II, 189 pages. 1967. DM 14,-

Vol. 40: J. Tits, Tabellen zu den einfachen Lie Gruppen und ihren Darstellungen. VI, 53 Seiten. 1967. DM 6,80

Vol. 41: A. Grothendieck, Local Cohomology. VI, 106 pages. 1967. DM 10,-

Vol. 42: J. F. Berglund and K. H. Hofmann, Compact Semitopological Semigroups and Weakly Almost Periodic Functions. VI, 160 pages. 1967. DM 12,-

Vol. 43: D. G. Quillen, Homotopical Algebra. VI, 157 pages. 1967. DM 14,-

Vol. 44: K. Urbanik, Lectures on Prediction Theory. IV, 50 pages. 1967. DM 5,80

Vol. 45: A. Wilansky, Topics in Functional Analysis. VI, 102 pages. 1967. DM 9,60

Vol. 46: P. E. Conner, Seminar on Periodic Maps. IV, 116 pages. 1967. DM 10,60

Vol. 47: Reports of the Midwest Category Seminar I. IV, 181 pages. 1967. DM 14,80

Vol. 48: G. de Rham, S. Maumary et M. A. Kervaire, Torsion et Type Simple d'Homotopie. IV, 101 pages. 1967. DM 9,60

Vol. 49: C. Faith, Lectures on Injective Modules and Quotient Rings. XVI, 140 pages. 1967. DM 12,80

Vol. 50: L. Zalcman, Analytic Capacity and Rational Approximation. VI, 155 pages. 1968. DM 13.20

Vol. 51: Séminaire de Probabilités II. IV, 199 pages. 1968. DM 14,-

Vol. 52: D. J. Simms, Lie Groups and Quantum Mechanics. IV, 90 pages. 1968. DM 8,-

Vol. 53: J. Cerf, Sur les difféomorphismes de la sphère de dimension trois ($\Gamma_4 = 0$). XII, 133 pages. 1968. DM 12,-

Vol. 54: G. Shimura, Automorphic Functions and Number Theory. VI, 69 pages. 1968. DM 8,-

Vol. 55: D. Gromoll, W. Klingenberg und W. Meyer, Riemannsche Geometrie im Großen. VI, 287 Seiten. 1968. DM 20,-

Vol. 56: K. Floret und J. Wloka, Einführung in die Theorie der lokalkonvexen Räume. VIII, 194 Seiten. 1968. DM 16,-

Vol. 57: F. Hirzebruch und K. H. Mayer, O (n)-Mannigfaltigkeiten, exotische Sphären und Singularitäten. IV, 132 Seiten. 1968. DM 10,80

Vol. 58: Kuramochi Boundaries of Riemann Surfaces. IV, 102 pages. 1968. DM 9,60

Vol. 59: K. Jänich, Differenzierbare G-Mannigfaltigkeiten. VI, 89 Seiten. 1968. DM 8,-

Vol. 60: Seminar on Differential Equations and Dynamical Systems. Edited by G. S. Jones. VI, 106 pages. 1968. DM 9,60

Vol. 61: Reports of the Midwest Category Seminar II. IV, 91 pages. 1968. DM 9,60

Vol. 62: Harish-Chandra, Automorphic Forms on Semisimple Lie Groups X, 138 pages. 1968. DM 14,-

Vol. 63: F. Albrecht, Topics in Control Theory. IV, 65 pages. 1968. DM 6,80

Vol. 64: H. Berens, Interpolationsmethoden zur Behandlung von Approximationsprozessen auf Banachräumen. VI, 90 Seiten. 1968. DM 8,-

Vol. 65: D. Kölzow, Differentiation von Maßen. XII, 102 Seiten. 1968. DM 8,-

Vol. 66: D. Ferus, Totale Absolutkrümmung in Differentialgeometrie und -topologie. VI, 85 Seiten. 1968. DM 8,-

Vol. 67: F. Kamber and P. Tondeur, Flat Manifolds. IV, 53 pages. 1968. DM 5,80

Vol. 68: N. Boboc et P. Mustață, Espaces harmoniques associés aux opérateurs différentiels linéaires du second ordre de type elliptique. VI, 95 pages. 1968. DM 8,60

Vol. 69: Seminar über Potentialtheorie. Herausgegeben von H. Bauer. VI, 180 Seiten. 1968. DM 14,80

Vol. 70: Proceedings of the Summer School in Logic. Edited by M. H. Löb. IV, 331 pages. 1968. DM 20,-

Vol. 71: Séminaire Pierre Lelong (Analyse), Année 1967 – 1968. VI, 190 pages. 1968. DM 14,

Vol. 72: The Syntax and Semantics of Infinitary Languages. Edited by J. Barwise. IV, 268 pages. 1968. DM 18,-

Vol. 73: P. E. Conner, Lectures on the Action of a Finite Group. IV, 123 pages. 1968. DM 10,-

Vol. 74: A. Fröhlich, Formal Groups. IV, 140 pages. 1968. DM 12,-

Vol. 75: G. Lumer, Algèbres de fonctions et espaces de Hardy. VI, 80 pages. 1968. DM 8,-

Vol. 76: R. G. Swan, Algebraic K-Theory. IV, 262 pages. 1968. DM 18,-

Vol. 77: P.-A. Meyer, Processus de Markov: la frontière de Martin. IV, 123 pages. 1968. DM 10,-

Vol. 78: H. Herrlich, Topologische Reflexionen und Coreflexionen. XVI, 166 Seiten. 1968. DM 12,-

Vol. 79: A. Grothendieck, Catégories Cofibrées Additives et Complexe Cotangent Relatif. IV, 167 pages. 1968. DM 12,-

Vol. 80: Seminar on Triples and Categorical Homology Theory. Edited by B. Eckmann. IV, 398 pages. 1969. DM 20,-

Vol. 81: J.-P. Eckmann et M. Guenin, Méthodes Algébriques en Mécanique Statistique. VI, 131 pages. 1969. DM 12,-

Vol. 82: J. Wloka, Grundräume und verallgemeinerte Funktionen. VIII, 131 Seiten. 1969. DM 12,-

Vol. 83: O. Zariski, An Introduction to the Theory of Algebraic Surfaces. IV, 100 pages. 1969. DM 8,-

Vol. 84: H. Lüneburg, Transitive Erweiterungen endlicher Permutationsgruppen. IV, 119 Seiten. 1969. DM 10,-

Vol. 85: P. Cartier et D. Foata, Problèmes combinatoires de commutation et réarrangements. IV, 88 pages. 1969. DM 8,-

Vol. 86: Category Theory, Homology Theory and their Applications I. Edited by P. Hilton. VI, 216 pages. 1969. DM 16,-

Vol. 87: M. Tierney, Categorical Constructions in Stable Homotopy Theory. IV, 65 pages. 1969. DM 6,-

Vol. 88: Séminaire de Probabilités III. IV, 229 pages. 1969. DM 18,-

Vol. 89: Probability and Information Theory. Edited by M. Behara, K. Krickeberg and J. Wolfowitz. IV, 256 pages. 1969. DM 18,-

Vol. 90: N. P. Bhatia and O. Hajek, Local Semi-Dynamical Systems. II, 157 pages. 1969. DM 14,-

Vol. 91: N. N. Janenko, Die Zwischenschrittmethode zur Lösung mehrdimensionaler Probleme der mathematischen Physik. VIII, 194 Seiten. 1969. DM 16,80

Vol. 92: Category Theory, Homology Theory and their Applications II. Edited by P. Hilton. V, 308 pages. 1969. DM 20,-

Vol. 93: K. R. Parthasarathy, Multipliers on Locally Compact Groups. III, 54 pages. 1969. DM 5,60

Vol. 94: M. Machover and J. Hirschfeld, Lectures on Non-Standard Analysis. VI, 79 pages. 1969. DM 6,-

Vol. 95: A. S. Troelstra, Principles of Intuitionism. II, 111 pages. 1969. DM 10,-

Vol. 96: H.-B. Brinkmann und D. Puppe, Abelsche und exakte Kategorien, Korrespondenzen. V, 141 Seiten. 1969. DM 10,-

Vol. 97: S. O. Chase and M. E. Sweedler, Hopf Algebras and Galois theory. II, 133 pages. 1969. DM 10,-

Vol. 98: M. Heins, Hardy Classes on Riemann Surfaces. III, 106 pages. 1969. DM 10,-

Vol. 99: Category Theory, Homology Theory and their Applications III. Edited by P. Hilton. IV, 489 pages. 1969. DM 24,-

Vol. 100: M. Artin and B. Mazur, Etale Homotopy. II, 196 Seiten. 1969. DM 12,-

Vol. 101: G. P. Szegö et G. Treccani, Semigruppi di Trasformazioni Multivoche. VI, 177 pages. 1969. DM 14,-

Vol. 102: F. Stummel, Rand- und Eigenwertaufgaben in Sobolewschen Räumen. VIII, 386 Seiten. 1969. DM 20,-

Vol. 103: Lectures in Modern Analysis and Applications I. Edited by C. T. Taam. VII, 162 pages. 1969. DM 12,-

Vol. 104: G. H. Pimbley, Jr., Eigenfunction Branches of Nonlinear Operators and their Bifurcations. II, 128 pages. 1969. DM 10,-

Vol. 105: R. Larsen, The Multiplier Problem. VII, 284 pages. 1969. DM 18,-

Vol. 106: Reports of the Midwest Category Seminar III. Edited by S. Mac Lane. III, 247 pages. 1969. DM 16,-

Vol. 107: A. Peyerimhoff, Lectures on Summability. III, 111 pages. 1969. DM 8,-

Vol. 108: Algebraic K-Theory and its Geometric Applications. Edited by R. M. F. Moss and C. B. Thomas. IV, 86 pages. 1969. DM 6,-

Vol. 109: Conference on the Numerical Solution of Differential Equations. Edited by J. Ll. Morris. VI, 275 pages. 1969. DM 18,-

Vol. 110: The Many Facets of Graph Theory. Edited by G. Chartrand and S. F. Kapoor. VIII, 290 pages. 1969. DM 18,-

Vol. 111: K. H. Mayer, Relationen zwischen charakteristischen Zahlen. III, 99 Seiten. 1969. DM 8,–

Vol. 112: Colloquium on Methods of Optimization. Edited by N. N. Moiseev. IV, 293 pages. 1970. DM 18,–

Vol. 113: R. Wille, Kongruenzklassengeometrien. III, 99 Seiten. 1970. DM 8,–

Vol. 114: H. Jacquet and R. P. Langlands, Automorphic Forms on GL (2) VII, 548 pages. 1970. DM 24,–

Vol. 115: K. H. Roggenkamp and V. Huber-Dyson, Lattices over Orders I XIX, 290 pages. 1970. DM 18,–

Vol. 116: Séminaire Pierre Lelong (Analyse) Année 1969. IV, 195 pages. 1970. DM 14,–

Vol. 117: Y. Meyer, Nombres de Pisot, Nombres de Salem et Analyse Harmonique. 63 pages. 1970. DM 6,–

Vol. 118: Proceedings of the 15th Scandinavian Congress, Oslo 1968. Edited by K. E. Aubert and W. Ljunggren. IV, 162 pages. 1970. DM 12,–

Vol. 119: M. Raynaud, Faisceaux amples sur les schémas en groupes et les espaces homogènes. III, 219 pages. 1970. DM 14,–

Vol. 120: D. Siefkes, Büchi's Monadic Second Order Successor Arithmetic. XII, 130 Seiten. 1970. DM 12,–

Vol. 121: H. S. Bear, Lectures on Gleason Parts. III, 47 pages. 1970. DM 6,–

Vol. 122: H. Zieschang, E. Vogt und H.-D. Coldewey, Flächen und ebene diskontinuierliche Gruppen. VIII, 203 Seiten. 1970. DM 16,–

Vol. 123: A. V. Jategaonkar, Left Principal Ideal Rings. VI, 145 pages. 1970. DM 12,–

Vol. 124: Séminare de Probabilités IV. Edited by P. A. Meyer. IV, 282 pages. 1970. DM 20,–

Vol. 125: Symposium on Automatic Demonstration. V, 310 pages. 1970 DM 20,–

Vol. 126: P. Schapira, Théorie des Hyperfonctions. XI, 157 pages. 1970. DM 14,–

Vol. 127: I. Stewart, Lie Algebras. IV, 97 pages. 1970. DM 10,–

Vol. 128: M. Takesaki, Tomita's Theory of Modular Hilbert Algebras and its Applications. II, 123 pages. 1970. DM 10,–

Vol. 129: K. H. Hofmann, The Duality of Compact Semigroups and C*-Bigebras. XII, 142 pages. 1970. DM 14,–

Vol. 130: F. Lorenz, Quadratische Formen über Körpern. II, 77 Seiten. 1970. DM 8,–

Vol. 131: A Borel et al., Seminar on Algebraic Groups and Related Finite Groups. VII, 321 pages. 1970. DM 22,–

Vol. 132: Symposium on Optimization. III, 348 pages. 1970. DM 22,–

Vol. 133: F. Topsøe, Topology and Measure. XIV, 79 pages. 1970. DM 8,–

Vol. 134: L. Smith, Lectures on the Eilenberg-Moore Spectral Sequence. VII, 142 pages. 1970. DM 14,–

Vol. 135: W. Stoll, Value Distribution of Holomorphic Maps into Compact Complex Manifolds. II, 267 pages. 1970. DM 18,–

Vol. 136: M. Karoubi et al., Séminaire Heidelberg-Saarbrücken-Strasbuorg sur la K-Théorie. IV, 264 pages. 1970. DM 18,–

Vol. 137: Reports of the Midwest Category Seminar IV. Edited by S. MacLane. III, 139 pages. 1970. DM 12,–

Vol. 138: D. Foata et M. Schützenberger, Théorie Géométrique des Polynômes Eulériens. V, 94 pages. 1970. DM 10,–

Vol. 139: A. Badrikian, Séminaire sur les Fonctions Aléatoires Linéaires et les Mesures Cylindriques. VII, 221 pages. 1970. DM 18,–

Vol. 140: Lectures in Modern Analysis and Applications II. Edited by C. T. Taam. VI, 119 pages. 1970. DM 10,–

Vol. 141: G. Jameson, Ordered Linear Spaces. XV, 194 pages. 1970. DM 16,–

Vol. 142: K. W. Roggenkamp, Lattices over Orders II. V, 388 pages. 1970. DM 22,–

Vol. 143: K. W. Gruenberg, Cohomological Topics in Group Theory. XIV, 275 pages. 1970. DM 20,–

Vol. 144: Seminar on Differential Equations and Dynamical Systems, II. Edited by J. A. Yorke. VIII, 268 pages. 1970. DM 20,–

Vol. 145: E. J. Dubuc, Kan Extensions in Enriched Category Theory. XVI, 173 pages. 1970. DM 16,–

Vol. 146: A. B. Altman and S. Kleiman, Introduction to Grothendieck Duality Theory. II, 192 pages. 1970. DM 18,–

Vol. 147: D. E. Dobbs, Cech Cohomological Dimensions for Commutative Rings. VI, 176 pages. 1970. DM 16,–

Vol. 148: R. Azencott, Espaces de Poisson des Groupes Localement Compacts. IX, 141 pages. 1970. DM 14,–

Vol. 149: R. G. Swan and E. G. Evans, K-Theory of Finite Groups and Orders. IV, 237 pages. 1970. DM 20,–

Vol. 150: Heyer, Dualität lokalkompakter Gruppen. XIII, 372 Seiten. 1970. DM 20,–

Vol. 151: M. Demazure et A. Grothendieck, Schémas en Groupes I. (SGA 3). XV, 562 pages. 1970. DM 24,–

Vol. 152: M. Demazure et A. Grothendieck, Schémas en Groupes II. (SGA 3). IX, 654 pages. 1970. DM 24,–

Vol. 153: M. Demazure et A. Grothendieck, Schémas en Groupes III. (SGA 3). VIII, 529 pages. 1970. DM 24,–

Vol. 154: A. Lascoux et M. Berger, Variétés Kähleriennes Compactes. VII, 83 pages. 1970. DM 8,–

Vol. 155: Several Complex Variables I, Maryland 1970. Edited by J. Horváth. IV, 214 pages. 1970. DM 18,–

Vol. 156: R. Hartshorne, Ample Subvarieties of Algebraic Varieties. XIV, 256 pages. 1970. DM 20,–

Vol. 157: T. tom Dieck, K. H. Kamps und D. Puppe, Homotopietheorie. VI, 265 Seiten. 1970. DM 20,–

Vol. 158: T. G. Ostrom, Finite Translation Planes. IV. 112 pages. 1970. DM 10,–

Vol. 159: R. Ansorge und R. Hass. Konvergenz von Differenzenverfahren für lineare und nichtlineare Anfangswertaufgaben. VIII, 145 Seiten. 1970. DM 14,–

Vol. 160: L. Sucheston, Constributions to Ergodic Theory and Probability. VII, 277 pages. 1970. DM 20,–

Vol. 161: J. Stasheff, H-Spaces from a Homotopy Point of View. VI, 95 pages. 1970. DM 10,–

Vol. 162: Harish-Chandra and van Dijk, Harmonic Analysis on Reductive p-adic Groups. IV, 125 pages. 1970. DM 12,–

Vol. 163: P. Deligne, Equations Différentielles à Points Singuliers Reguliers. III, 133 pages. 1970. DM 12,–

Vol. 164: J. P. Ferrier, Seminaire sur les Algebres Complètes. II, 69 pages. 1970. DM 8,–

Vol. 165: J. M. Cohen, Stable Homotopy. V, 194 pages. 1970. DM 16,–

Vol. 166: A. J. Silberger, PGL_2 over the p-adics: its Representations, Spherical Functions, and Fourier Analysis. VII, 202 pages. 1970. DM 18,–

Vol. 167: Lavrentiev, Romanov and Vasiliev, Multidimensional Inverse Problems for Differential Equations. V, 59 pages. 1970. DM 10,–

Vol. 168: F. P. Peterson, The Steenrod Algebra and its Applications: A conference to Celebrate N. E. Steenrod's Sixtieth Birthday. VII, 317 pages. 1970. DM 22,–

Vol. 169: M. Raynaud, Anneaux Locaux Henséliens. V, 129 pages. 1970. DM 12,–

Vol. 170: Lectures in Modern Analysis and Applications III. Edited by C. T. Taam. VI, 213 pages. 1970. DM 18,–

Vol. 171: Set-Valued Mappings, Selections and Topological Properties of 2^X. Edited by W. M. Fleischman. X, 110 pages. 1970. DM 12,–

Vol. 172: Y.-T. Siu and G. Trautmann, Gap-Sheaves and Extension of Coherent Analytic Subsheaves. V, 172 pages. 1971. DM 16,–

Vol. 173: J. N. Mordeson and B. Vinograde, Structure of Arbitrary Purely Inseparable Extension Fields. IV, 138 pages. 1970. DM 14,–

Vol. 174: B. Iversen, Linear Determinants with Applications to the Picard Scheme of a Family of Algebraic Curves. VI, 69 pages. 1970. DM 8,–

Vol. 175: M. Brelot, On Topologies and Boundaries in Potential Theory. VI, 176 pages. 1971. DM 18,–

Vol. 176: H. Popp, Fundamentalgruppen algebraischer Mannigfaltigkeiten. IV, 154 Seiten. 1970. DM 16,–

Vol. 177: J. Lambek, Torsion Theories, Additive Semantics and Rings of Quotients. VI, 94 pages. 1971. DM 12,–

Vol. 178: Th. Bröcker und T. tom Dieck, Kobordismentheorie. XVI, 191 Seiten. 1970. DM 18,–

Vol. 179: Seminaire Bourbaki – vol. 1968/69. Exposés 347-363. IV. 295 pages. 1971. DM 22,–

Vol. 180: Séminaire Bourbaki – vol. 1969/70. Exposés 364-381. IV, 310 pages. 1971. DM 22,–

Vol. 181: F. DeMeyer and E. Ingraham, Separable Algebras over Commutative Rings. V, 157 pages. 1971. DM 16,–